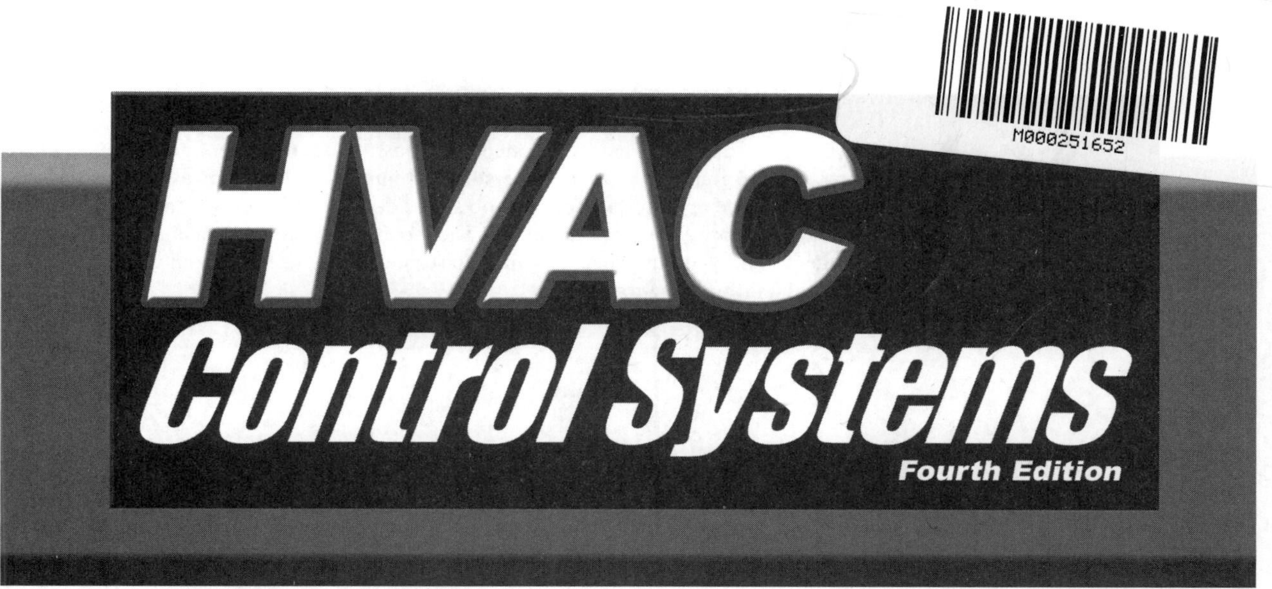

Workbook

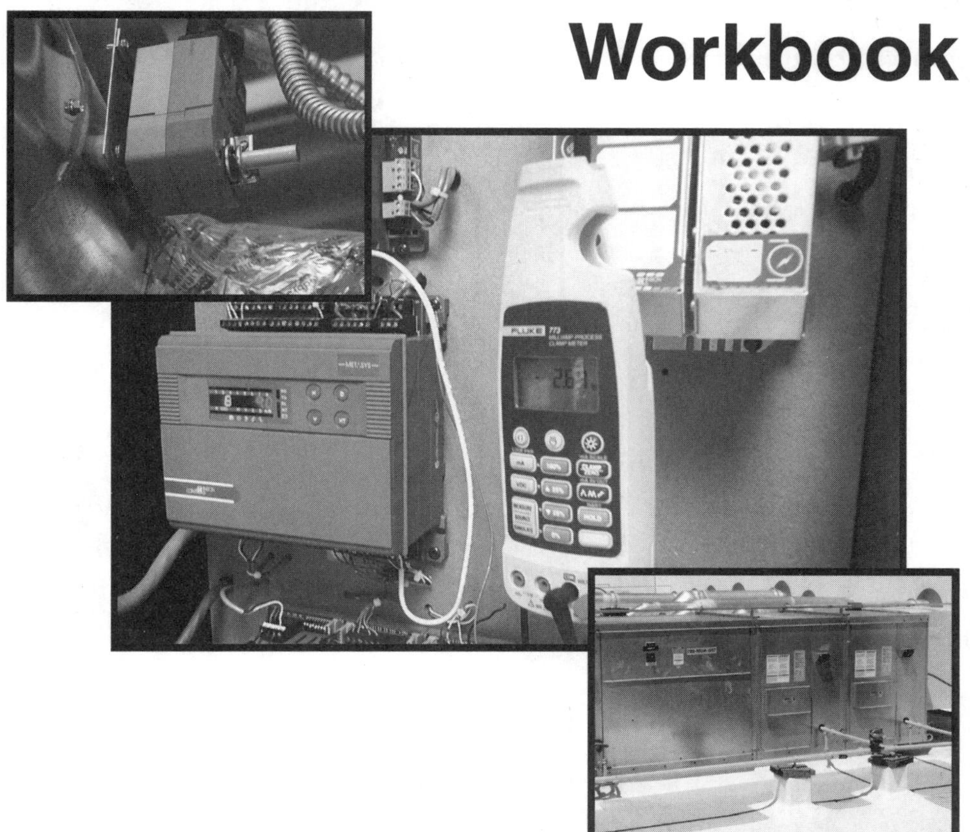

Ronnie J. Auvil

The publisher is grateful for the technical information and assistance provided by the following companies and organizations:

Jackson Systems, LLC
Spirax Sarco, Inc.

4 5 6 7 8 9 – 17 – 9 8 7 6 5 4 3

Printed in the United States of America

ISBN 978-0-8269-0780-6

 This book is printed on recycled paper.

CONTENTS

Section 1.1 Comfort

REVIEW QUESTIONS

Name ______________________________ **Date** ________________

Multiple Choice

______________ **1.** ___ is the process that occurs when a liquid changes to a vapor by absorbing heat.
A. Ventilation
B. Conduction
C. Radiation
D. Evaporation

______________ **2.** Comfort is usually attained at normal cooling and heating temperatures with a humidity level of about ___.
A. 35%
B. 50%
C. 65%
D. 70%

______________ **3.** Air in a building must be circulated ___ to provide maximum comfort.
A. at a low velocity
B. at a high velocity
C. intermittently
D. continuously

______________ **4.** The five requirements for comfort are proper ___.
A. heat transfer, circulation, filtration, humidity, and physiological function
B. temperature, humidity, filtration, circulation, and ventilation
C. heat transfer, filtration, psychrometrics, circulation, and enthalpy
D. temperature, dehumidification, filtration, circulation, and heat transfer

______________ **5.** ___ air is air that is used to replace air that is lost to exhaust.
A. Makeup
B. Replacement
C. Moist
D. Atmospheric

Completion

______________ **1.** ___ is the condition that occurs when people cannot sense a difference between themselves and the surrounding air.

______________ **2.** ___ is the temperature of air below which moisture begins to condense from the air.

______________ 3. ___ is the process of removing particulate matter from air that circulates through an air distribution system.

______________ 4. ___ is the process of introducing fresh air into a building.

______________ 5. ___ is the measurement of the intensity of the heat in a substance.

Circulation

______________ 1. filter

______________ 2. circulating air

______________ 3. supply air

______________ 4. return air grill

______________ 5. outside air in

______________ 6. return duct

______________ 7. heating coil

______________ 8. return air

______________ 9. supply air fan

______________ 10. supply duct

______________ 11. supply air register

______________ 12. cooling coil

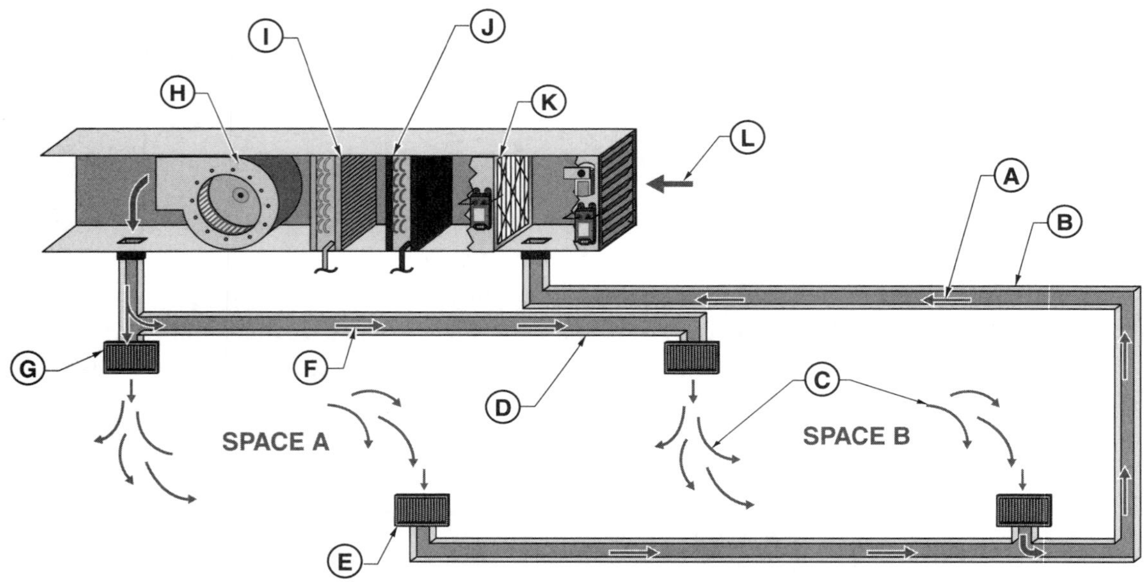

Section 1.2 Thermodynamics

REVIEW QUESTIONS

Name ______________________________ **Date** ________________

Multiple Choice

________________ **1.** A British thermal unit (Btu) is the amount of heat energy required to raise the temperature of 1 lb of water ___.

A. 1°F
B. 100°F
C. to desired temperature
D. in 24 hr

________________ **2.** ___ is the force with which a body is pulled downward by gravity.

A. Weight
B. Density
C. Viscosity
D. Mass

________________ **3.** ___ is the amount of heat required to melt a ton of ice (2000 lb) over a 24-hour period.

A. A British thermal unit (Btu)
B. Latent heat
C. Specific heat
D. A ton of cooling

________________ **4.** The specific heat of air is ___ Btu/lb/°F, which means that 1 lb of water holds as much heat energy as approximately 4 lb of air.

A. 0.24
B. 0.75
C. 1.25
D. 4.00

________________ **5.** ___ is expressed as the ratio of the quantity of heat required to raise the temperature of a material 1°F compared to the quantity required to raise the temperature of an equal mass of water 1°F.

A. Sensible heat
B. Specific heat
C. Latent heat
D. Warmth

_______________ **6.** The ___ states that heat always flows from a material at a high temperature to a material at a low temperature.
- A. first law of thermodynamics
- B. second law of thermodynamics
- C. third law of thermodynamics
- D. law of psychrometrics

Completion

_______________ **1.** ___ is heat identified by a change of state and no temperature change.

_______________ **2.** ___ is heat transfer that occurs when molecules in a material are heated and the heat is passed from molecule to molecule through the material.

_______________ **3.** ___ is heat transfer that occurs when currents circulate between warm and cool regions of a fluid.

_______________ **4.** ___ is heat transfer in the form of radiant energy (electromagnetic waves).

_______________ **5.** ___ is the science of thermal energy (heat) and how it transforms to and from other forms of energy.

_______________ **6.** Heat is transferred when the radiant energy waves contact a(n) ___ object.

_______________ **7.** ___ is the measurement of energy contained in a substance and is identified by a temperature difference or a change of state.

Heat Transfer

_______________ **1.** convection

_______________ **2.** radiation

_______________ **3.** conduction

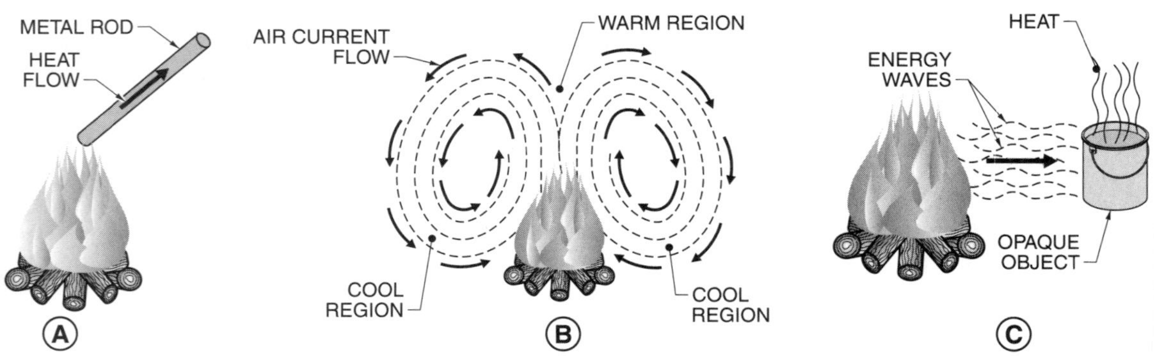

Section 1.3 Psychrometrics

REVIEW QUESTIONS

Name ______________________________ Date ____________________

Multiple Choice

______________ **1.** ___ is the elements that make up atmospheric air with moisture and particles removed.
A. Standard air
B. Dry air
C. Free air
D. Exhaust air

______________ **2.** Absolute pressure is pressure above ___.
A. atmospheric pressure
B. gauge pressure
C. a water column
D. a perfect vacuum

______________ **3.** ___ is the scientific study of the properties of air and the relationships between them.
A. Psychrometrics
B. Combustion
C. Thermodynamics
D. Stratification

______________ **4.** A psychrometric chart defines ___.
A. the properties of the air at various conditions
B. indoor air quality ratings
C. enthalpy
D. the difference between the Fahrenheit and the Celsius scales

Completion

______________ **1.** ___ is the temperature of the air without reference to the humidity level.

______________ **2.** Wet bulb temperature is measured using a(n) ___.

______________ **3.** ___ is the force created by a substance per unit of area.

______________ **4.** ___ is the most important variable that is measured and controlled in a commercial HVAC system.

______________ **5.** ___ is the total heat contained in a material.

______________ **6.** ___ is the most important variable that is measured and controlled in a commercial HVAC system.

______________ **7.** ___ is the amount of moisture in the air compared to the amount of moisture that it could hold if it were saturated (full of water).

Calculating Pressure

_______________ **1.** Box A pressure = ___ psi.

_______________ **2.** Box B pressure = ___ psi.

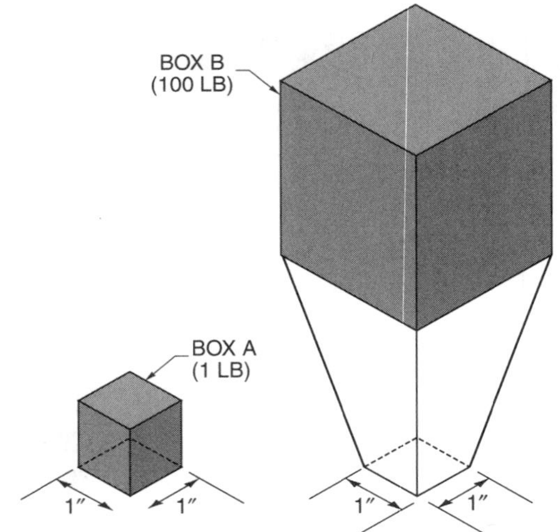

Fahrenheit to Celsius Temperature Conversion

_______________ **1.** A temperature of 95°F = ___°C.

_______________ **2.** A temperature of 140°F = ___°C.

_______________ **3.** A temperature of 55°F = ___°C.

_______________ **4.** A temperature of 74°F = ___°C.

_______________ **5.** A temperature of 68°F = ___°C.

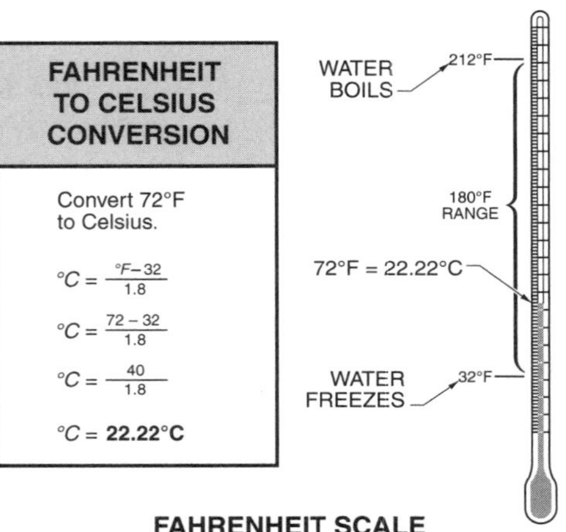

Celsius to Fahrenheit Temperature Conversion

_______________ **1.** A temperature of 20°C = ___°F.

_______________ **2.** A temperature of 65°C = ___°F.

_______________ **3.** A temperature of 10°C = ___°F.

_______________ **4.** A temperature of 27°C = ___°F.

_______________ **5.** A temperature of 40°C = ___°F.

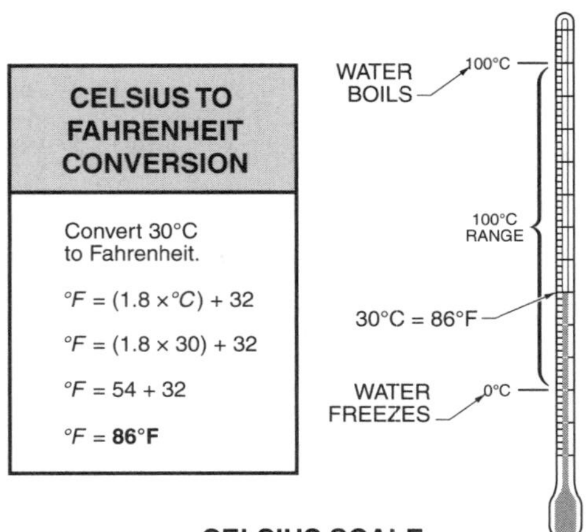

HVAC Fundamentals

ACTIVITIES

Name ______________________________ Date ____________________

Activity 1-1. Psychrometrics

A complaint is received that a room is too hot. It is 7 AM in the summer and work has just started. The room conditions are checked with an electronic hand-held temperature and humidity meter. The temperature is 85°F and the humidity is 55% rh.

PSYCHROMETRIC CHART A

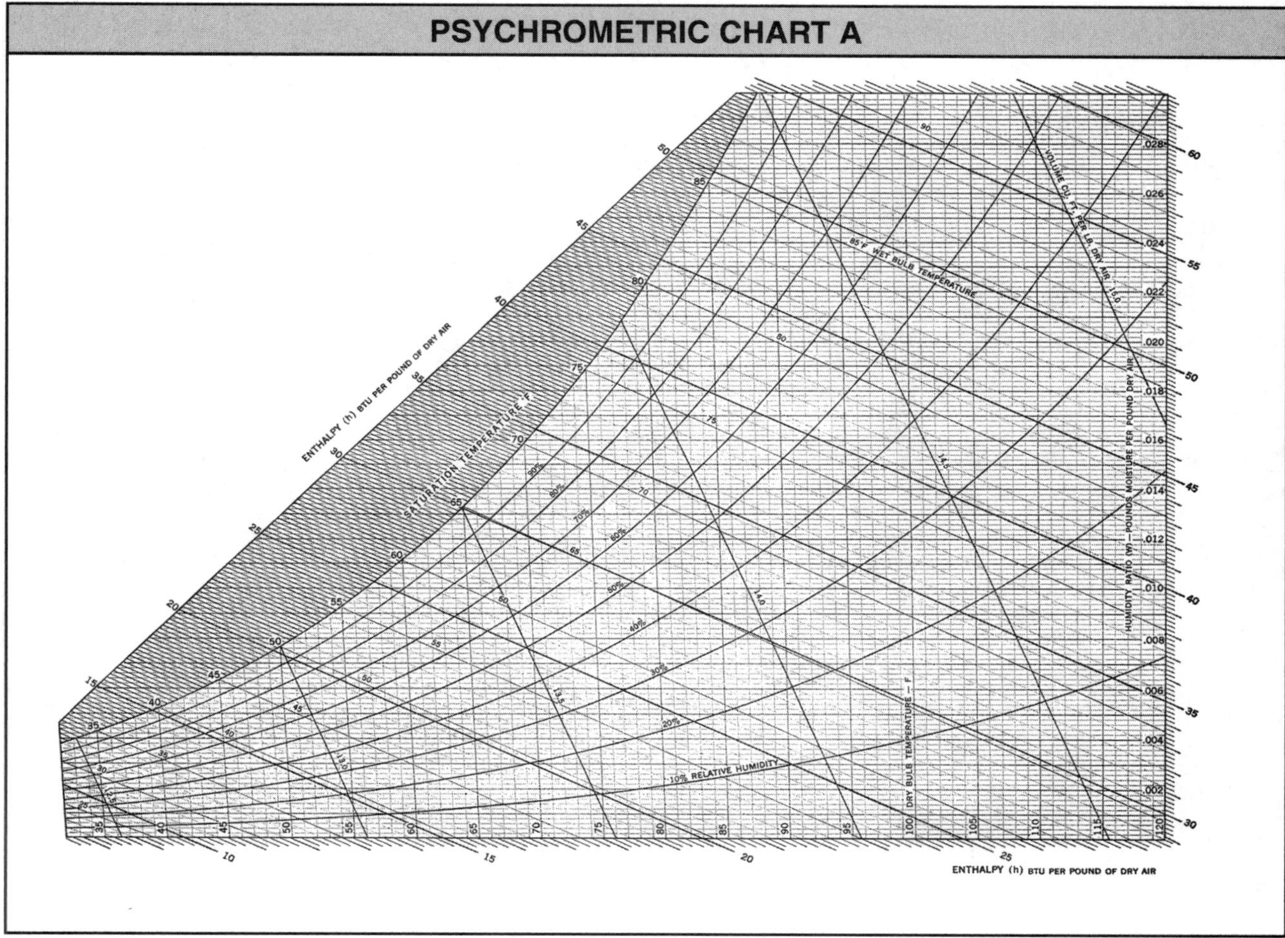

1. Plot this point on psychrometric chart A.

____________________ **2.** Is this point outside the normal comfort zone?

After checking the building automation system workstation computer, it was found that the time schedule was incorrect, keeping the cooling unit OFF. The cooling unit is switched ON, and after 10 min the discharge air conditions are checked in the room. The temperature is 72°F and the humidity is 45% rh.

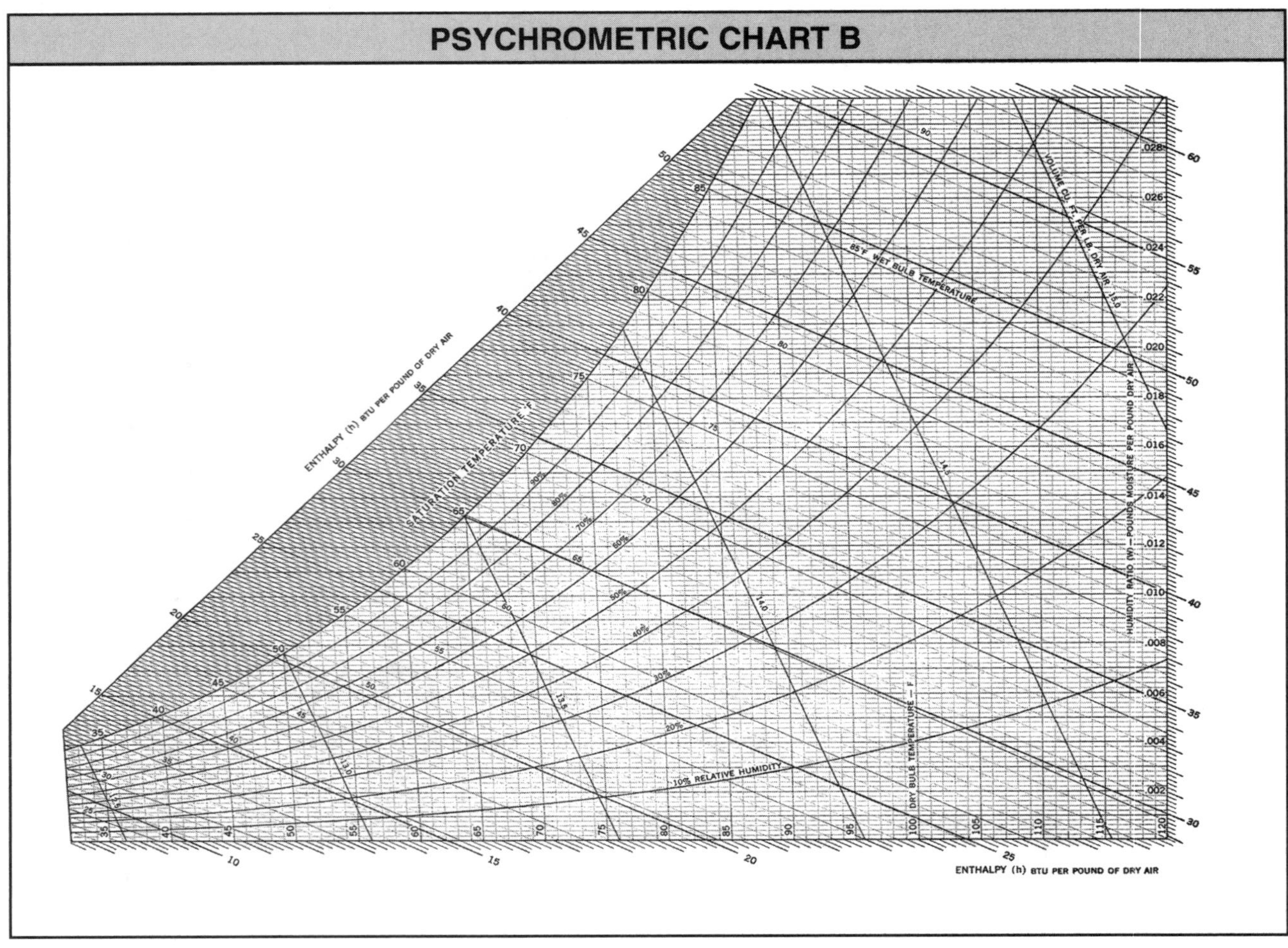

3. Plot this point on psychrometric chart B.

____________________ **4.** Is this point outside the normal comfort zone?

To monitor systems operation, a data trend is set up to monitor the room temperature and humidity every 15 min. The trend begins at 8 AM and ends at 10 AM. At 10 AM the data is checked.

ROOM TEMPERATURE AND HUMIDITY DATA TREND

TIME	TEMPERATURE*	HUMIDITY†
8:15 AM	85	55
8:30 AM	83	55
8:45 AM	81	52
9:00 AM	80	52
9:15 AM	78	48
9:30 AM	76	45
9:45 AM	75	42
10:00 AM	74	41

* in °F
† in % rh

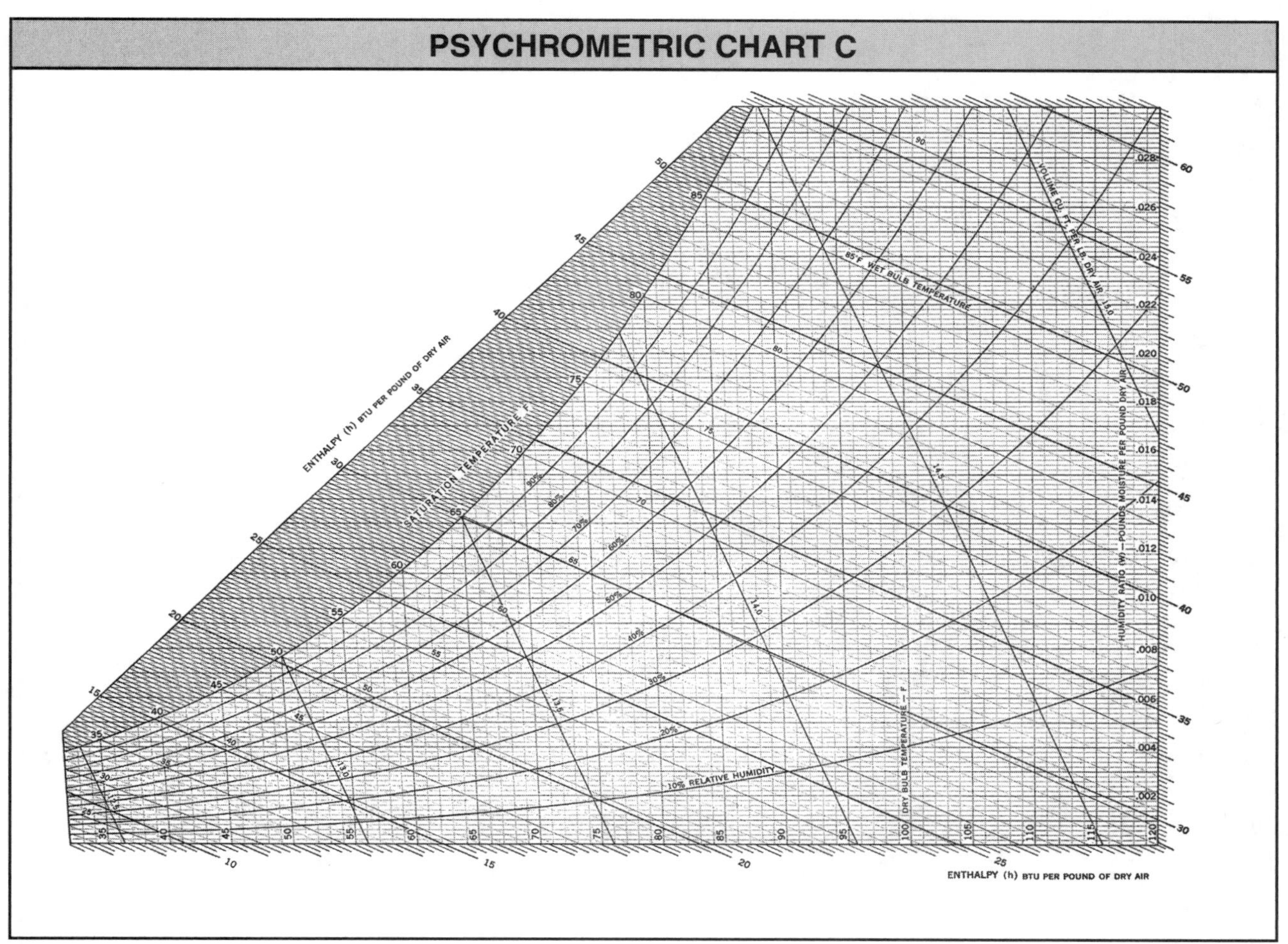

5. Plot the points as closely as possible on psychrometric chart C, then connect them in a straight line.

____________________ **6.** Are the conditions at 10 AM outside the normal comfort zone?

Section 2.1 Steam and Hot Water Heating Systems

REVIEW QUESTIONS

Name ______________________ **Date** ______________

Multiple Choice

__________ **1.** ___ from a high pressure boiler is commonly used in process heating, laundering, sterilizing, and many other applications.
- A. Cold water
- B. Hot water
- C. Dew
- D. Steam

__________ **2.** ___ can be classified as steam or hot water, by pressure, by heat production method, by fuel source for combustion, by design, or by construction materials.
- A. Air handling units
- B. Boilers
- C. Condensers
- D. Dampers

__________ **3.** A(n) ___ is a component directly attached to a boiler that is required for the operation of the boiler.
- A. fitting
- B. accessory
- C. terminal unit
- D. fuel oil strainer

__________ **4.** A(n) ___ opens in proportion to the amount of overpressure.
- A. temperature-pressure gauge
- B. aquastat
- C. vent
- D. relief valve

__________ **5.** Common steam boiler ___ include feedwater pumps, feedwater heaters, feedwater regulators, and steam traps.
- A. fittings
- B. controls
- C. accessories
- D. components

__________ **6.** A ___ valve is a hot water boiler accessory that reduces incoming makeup water pressure to 12 psi.
- A. relief
- B. flow control
- C. pressure-reducing
- D. control

______________ **7.** A(n) ___ prevents air from being trapped inside the boiler and causing excessive pressure buildup.
A. boiler vent
B. relief valve
C. aquastat
D. safety valve

______________ **8.** A low pressure steam heating boiler is a boiler operated at pressures of not more than ___ psi.
A. 15
B. 50
C. 100
D. 160

______________ **9.** A ___ boiler is a boiler in which steam or other vapor is generated at pressures of more than 15 psi.
A. power hot water
B. hot water supply
C. power steam
D. small power

______________ **10.** The ___ is located at the highest part of the steam side of the boiler.
A. surface blowdown valve
B. pressure control
C. safety valve
D. steam outlet

Completion

______________ **1.** A(n) ___ is a closed metal container (pressure vessel) in which water is heated to produce steam or hot water.

______________ **2.** A(n) ___ boiler is a boiler used for heating water or liquid to pressures of more than 160 psi or to temperatures of more than 205°F.

______________ **3.** A(n) ___ gauge is a boiler fitting that displays the amount of pressure in pounds per square inch (psi) inside a boiler.

______________ **4.** A(n) ___ gauge is a pressure gauge that indicates vacuum in in. Hg and pressure in psi on the same gauge.

______________ **5.** A(n) ___ is a boiler fitting that reduces the turbulence of boiler water to provide an accurate water level in the gauge glass.

______________ **6.** A(n) ___ is a hot water boiler fitting that measures the water temperature in the boiler and controls the temperature by starting and stopping the boiler burner.

______________ **7.** A(n) ___ is a boiler accessory that maintains the NOWL in the boiler by controlling the amount of feedwater pumped to the boiler from the surge tank.

_______________ 8. A(n) ___ system is a boiler system that provides fuel for combustion to produce the necessary heat in a boiler.

_______________ 9. A(n) ___ system is a boiler system that regulates the flow of air to and from the burner.

_______________ 10. ___ draft is draft produced using a mechanical device, such as a fan or blower.

_______________ 11. The surface blowdown line is piping located at the ___ of a boiler.

_______________ 12. The ASME Code states that a try lever test should be performed on a relief valve every ___ that the boiler is in operation or after any period of inactivity.

_______________ 13. A(n) ___ is a hot water heating system accessory that moves water through the boiler, system piping, and heating units.

_______________ 14. The ___ is the most important fitting on a steam boiler.

_______________ 15. ___ is the process of discharging water and undesirable accumulated material from a boiler.

_______________ 16. A temperature-pressure gauge has ___ scale(s).

_______________ 17. Draft may be natural or ___.

_______________ 18. A(n) ___ is a piece of equipment that is not directly attached to a boiler but is required for the operation of the boiler.

_______________ 19. A ___ control is a control that regulates the fuel supply, air supply, air-to-fuel ratio, and removal of gases of combustion in a boiler.

_______________ 20. A(n) ___ is a boiler fitting that reduces the turbulence of boiler water to provide an accurate water level in the gauge glass.

Safety Valves

_______________ 1. inlet

_______________ 2. spindle

_______________ 3. valve seat

_______________ 4. valve disc

_______________ 5. try lever

_______________ 6. discharge

_______________ 7. spring

_______________ 8. valve body

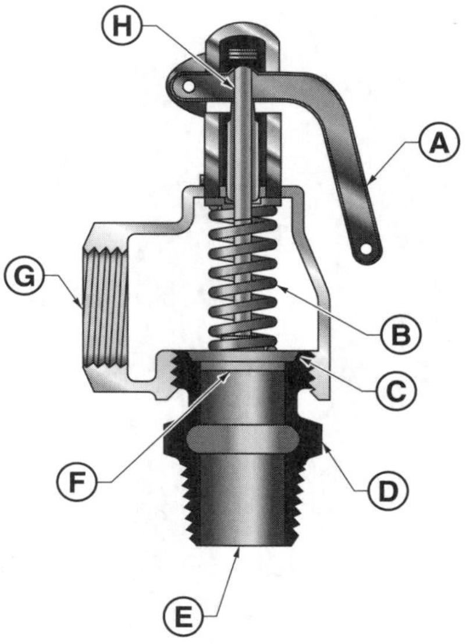

Pressure Gauges

_______________ **1.** vacuum

_______________ **2.** compound

_______________ **3.** steam pressure

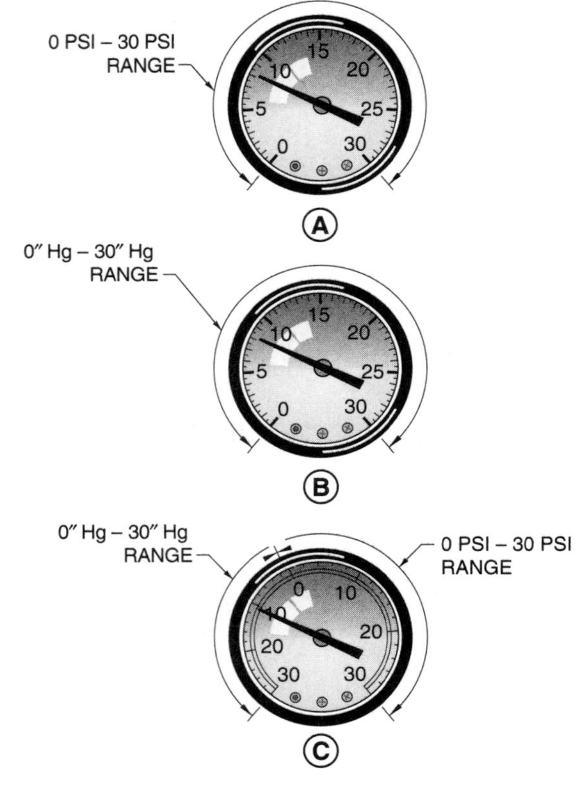

Heating Systems

_______________ **1.** piping system

_______________ **2.** steam header

_______________ **3.** controls

_______________ **4.** steam trap

_______________ **5.** steam boiler

_______________ **6.** heating unit

_______________ **7.** main steam line

_______________ **8.** feedwater pump

_______________ **9.** main steam stop valve

_______________ **10.** condensate return line

_______________ **11.** condensate receiver tank

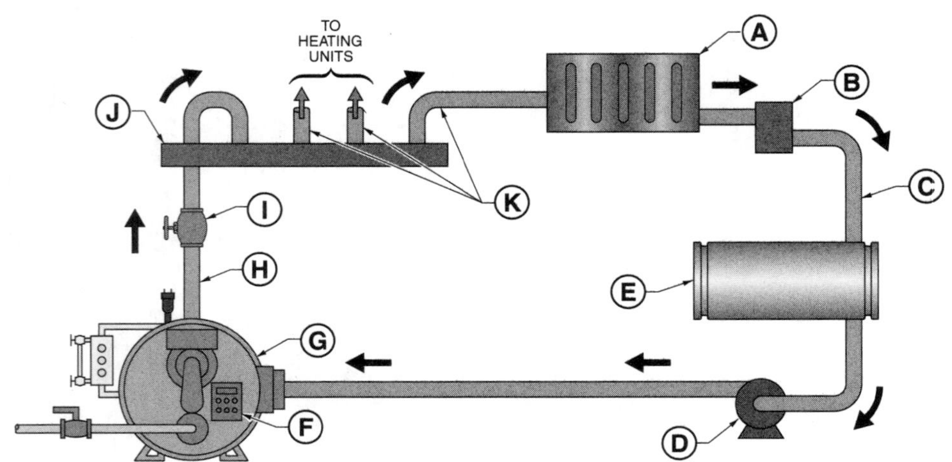

Section 2.2 Other Commercial Heating Systems

REVIEW QUESTIONS

Name ______________________________ **Date** ____________________

Multiple Choice

__________ **1.** A(n) ___ is a heating system that uses electric resistance heating elements and is located along the base of the outside walls of a building.
- A. electric radiant heat panel
- B. electric baseboard heater
- C. electric heating element
- D. solar panel

__________ **2.** An advantage of an electric radiant heat panel is that ___.
- A. there is no possibility of electrical shock
- B. electric heat costs less per Btu than natural gas and fossil fuels
- C. the air conditioning function is built-in
- D. no blower or air filters are necessary

__________ **3.** Water-to-air heat pumps have a ___ as the heat-transfer medium.
- A. coolant solution
- B. cooling coil containing refrigerant
- C. coil heat exchanger with water
- D. refrigerant line

__________ **4.** In solar hot water heating systems, ___ collectors are used when a high temperature or a large amount of water is required.
- A. evacuated-tube
- B. hot water
- C. vacuum
- D. flat-plate

Completion

__________ **1.** A(n) ___ is a direct expansion refrigeration system that contains devices and controls that reverse the flow of refrigerant.

__________ **2.** A(n) ___ is a ceiling panel with an embedded electric resistance heating element.

__________ **3.** The heat load calculation indicates the amount of heat a building space ___ in Btu per hour (Btu/hr).

__________ **4.** ___ heating technologies collect thermal energy from the sun and use this heat to provide hot water and building space heating for residential, commercial, and industrial applications.

__________ **5.** The heat load calculation indicates the amount of heat a building space loses in ___.

Rooftop Packaged Units

_____________________ **1.** condenser

_____________________ **2.** natural gas supply

_____________________ **3.** filter section

_____________________ **4.** exhaust hood

_____________________ **5.** compressor

_____________________ **6.** return air

_____________________ **7.** heat exchanger

_____________________ **8.** cooling section

_____________________ **9.** supply air fan

_____________________ **10.** evaporator coil

_____________________ **11.** supply plenum

_____________________ **12.** return air fan

_____________________ **13.** supply air to building space

_____________________ **14.** outside air louvers

_____________________ **15.** outside air

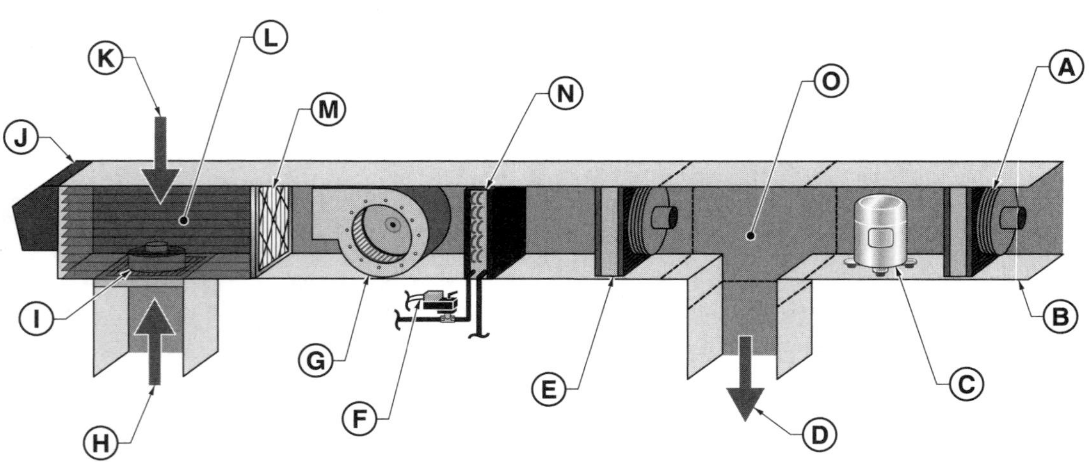

ACTIVITIES

Name ______________________________ Date ____________________

Activity 2-1. Boiler Nameplate

Use the boiler nameplate to answer the questions.

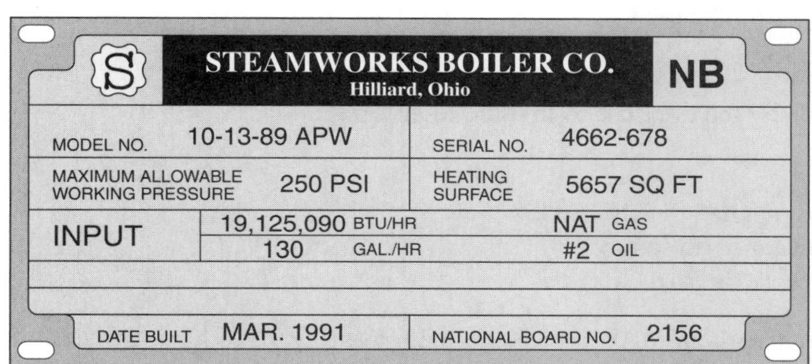

________________ **1.** The boiler manufacturer is ___.

________________ **2.** The boiler model number is ___.

________________ **3.** The boiler serial number is ___.

________________ **4.** The boiler burner rating is ___ Btu/hr.

________________ **5.** The listed heating surface area of the boiler is ___ sq ft.

________________ **6.** The boiler is designed to use ___ gas fuel.

________________ **7.** What weight oil can be used?

________________ **8.** How many gallons of oil are used per hour?

Activity 2-2. Safety Valves

In a safety valve, a spring exerts a downward force to keep the valve closed. The steam pressure exerts an upward force on the safety valve disc. The safety valve opens when the total force exerted by the steam exceeds the force exerted by the spring. The total force of the steam against the valve disc is equal to the area of the safety valve disc multiplied by the steam pressure. To find the total force applied to a safety valve, apply the following procedure:

1. Find the safety valve disc area. Valve disc area is calculated by applying the formula:

$A = \pi/4 \times d^2$

where

A = area of valve disc (in sq in.)

d = diameter of valve disc (in in.)

For example, the area of a valve with a disc diameter of 2½″ is 4.91 sq in.

$A = \pi/4 \times d^2$

$A = \pi/4 \times 2.52^2$

$A = 0.7854 \times 6.25$

$A = \mathbf{4.9087\ sq\ in.}$

2. Multiply the area by the pressure to find the total force.

$TF = A \times P$

where

TF = total force applied to valve (in lb)

A = area of valve disc (in sq in.)

P = steam pressure (in psi)

For example, a boiler with an MAWP of 60 psi with a safety valve having a 2½″ valve disc has a maximum total force of 294 lb.

$TF = A \times P$

$TF = 4.9087 \times 60$

$TF = \mathbf{294.522\ lb}$

A boiler has an MAWP of 80 psi. The boiler has two safety valves that each have a disc diameter of 3″.

______________________ **1.** What is the area of each valve disc?

______________________ **2.** What is the total area of the safety valves exposed to the steam pressure?

______________________ **3.** What is the maximum total force exerted by the steam against each valve disc?

Section 3.1 Types of Cooling Systems

REVIEW QUESTIONS

Name ______________________________ Date ______________

Multiple Choice

______________ **1.** A(n) ___ economizer is an economizer that operates strictly in response to the outside air temperature with no reference to the humidity values of the air.
A. wet bulb
B. dry bulb
C. enthalpy
D. centrifugal

______________ **2.** A ___ is a mechanical device that compresses refrigerant or other fluid.
A. compressor
B. chiller
C. centrifugal fan
D. cooling tower

______________ **3.** A ___ is a heat exchanger that removes heat from high-pressure refrigerant vapor.
A. compressor
B. chiller
C. centrifugal fan
D. condenser

______________ **4.** A(n) ___ is a heat exchanger that adds heat to low-pressure liquid refrigerant.
A. cooling tower
B. chiller
C. expansion device
D. evaporator

______________ **5.** A ___ is a device that cools water.
A. chiller
B. cooling tower
C. centrifugal fan
D. compressor

______________ **6.** A(n) ___ is an absorption refrigeration system component that vaporizes and separates the refrigerant from the absorbant.
A. economizer
B. generator
C. impeller
D. refrigerant pump

Completion

_______________ **1.** A(n) ___ is an HVAC system that uses outside air for cooling.

_______________ **2.** ___ is the total heat contained in a material.

_______________ **3.** A(n) ___ refrigeration system is a refrigeration system in which cooling is produced by the vaporization of refrigerant in a closed system.

_______________ **4.** A(n) ___ compressor is a compressor that uses one or more pistons that reciprocate inside closed cylinders to compress refrigerant vapor.

_______________ **5.** A(n) ___ is a valve or mechanical device that reduces the pressure on a liquid refrigerant by allowing the refrigerant to expand.

_______________ **6.** A(n) ___ refrigeration system is a refrigeration system that uses the absorption of one chemical by another chemical and heat transfer to produce a refrigeration effect.

_______________ **7.** A(n) ___ is an economizer that uses temperature and humidity levels of the outside air to control its operation.

_______________ **8.** ___ heat pumps are commonly water-to-air heat pumps.

_______________ **9.** Chillers include metering devices, evaporators, condensers, and compressors that are similar to ___.

_______________ **10.** A(n) ___ compressor is a compressor that uses two matching scrolls to compress refrigerant vapor.

_______________ **11.** A(n) ___ is a fluid that has a lower temperature than the refrigerant, which causes heat to flow to the medium.

_______________ **12.** The ___ is the device in an absorption system in which the refrigerant is absorbed by an absorbant.

Section 3.2 Cooling Towers

REVIEW QUESTIONS

Name ______________________________ **Date** ______________

Multiple Choice

_______________ **1.** A(n) ___ cooling tower is a cooling tower in which the fans are located in the airstream entering the tower in order to push the air through.
- A. crossflow
- B. counterflow
- C. forced-draft
- D. induced-draft

_______________ **2.** ___ fans are able to move large quantities of air at the low static pressures required in cooling towers.
- A. Propeller
- B. Variable-pitch
- C. Centrifugal
- D. Variable-speed

_______________ **3.** Cooling towers are categorized as either forced-draft or ___-draft.
- A. mechanical
- B. induced
- C. natural
- D. counterflow

_______________ **4.** Cooling towers that require fans approximately ___″ and larger in diameter are normally equipped with variable-pitch fans.
- A. 5
- B. 19
- C. 48
- D. 74

_______________ **5.** ___ are used to control, regulate, isolate, or modulate flow through the water lines serving a cooling tower.
- A. Valves
- B. Generators
- C. Condensers
- D. Fans

Completion

_______________ **1.** A(n) ___ condenser is a condenser that uses water as the condensing medium.

_______________ **2.** A(n) ___ cooling tower is a cooling tower in which the air moves vertically upward through the fill, counter to the downward fall of water.

______________ **3.** ___ valves are used to automatically replenish the normal water losses from the system.

______________ **4.** A(n) ___ cooling tower is a cooling tower in which the fans are located in the airstream leaving the tower in order to pull the air through.

______________ **5.** Cooling tower fan speed is commonly controlled by a(n) ___.

Cooling Towers

______________ **1.** cossover connection

______________ **2.** shell

______________ **3.** makeup water

______________ **4.** cooling tower

______________ **5.** warm water

______________ **6.** cool water in

______________ **7.** water pump

______________ **8.** refrigerant out

______________ **9.** tubes

______________ **10.** refrigerant in

______________ **11.** sump

______________ **12.** water-cooled condenser

______________ **13.** fill material

______________ **14.** warm air out

______________ **15.** warm water out

______________ **16.** air in

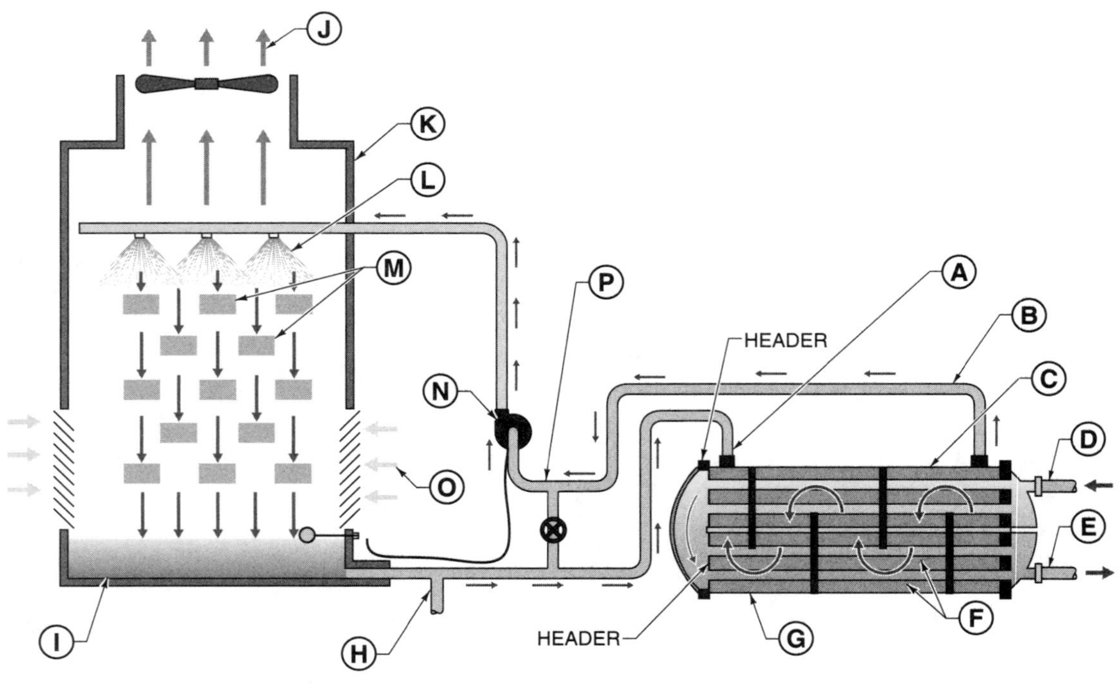

Section 3.3 Water Distribution Systems

REVIEW QUESTIONS

Name ______________________________ Date ____________________

Multiple Choice

______________ **1.** A(n) ___ system is a piping system in which no opening to the atmosphere exists.
A. closed
B. open
C. two-pipe
D. primary loop

______________ **2.** In a(n) ___ piping system, one or more chilled water pumps are controlled to supply chilled water to the building terminal units.
A. closed
B. open
C. two-pipe
D. primary loop

______________ **3.** Two-pipe systems consist of a ___.
A. separate supply line and return line
B. direct-return line and reverse-return line
C. chilled water supply line and a hot water return line
D. reverse-return pipe and a supply line

______________ **4.** ___ may be alleviated by the use of a reverse-return two-pipe system.
A. Pressure drops
B. Flow imbalances
C. Mechanical stresses
D. System failures

______________ **5.** In a primary loop piping system variable-speed pumps are common on the secondary loop to ___.
A. keep the water pumping through the system
B. add water into the system to make up for lost water
C. regulate flow imbalances
D. accommodate load changes

______________ **6.** A(n) ___ system is a piping system in which no opening to the atmosphere exists.
A. underground
B. vacuum
C. closed
D. separate

_______________ **7.** One method used to perform pressure control is through the use of ___.
A. bypass valves
B. return lines
C. expansion devices
D. compressors

_______________ **8.** An advantage of the primary loop piping system is the ___..
A. pressure capacity
B. efficiency
C. variable controls
D. simplicity

Completion

_______________ **1.** A(n) ___ system is a piping system in which the piping is open to the atmosphere at some point.

_______________ **2.** In a ___ piping system, one or more chilled water pumps are controlled to supply chilled water to the building terminal units.

_______________ **3.** Three-pipe systems have a single common ___ used for hot and chilled water.

_______________ **4.** Piping system designs include two-pipe, three-pipe, and ___ systems.

_______________ **5.** In a direct-return two-pipe system, the terminal unit closest to the ___ has the shortest supply and return lines.

Commercial Building Cooling Systems

ACTIVITIES

Name ______________________ **Date** ______________

Activity 3-1. Maintenance Checklist

Use the maintenance checklist to answer the questions.

MAINTENANCE CHECKLIST — COOLING TOWERS													
ONCE EVERY MONTH		JAN	FEB	MAR	APR	MAY	JUN	JUL	AUG	SEPT	OCT	NOV	DEC
	Date:	1/4	2/10										
1. Check fan and motor bearings and lubricate if necessary.		RA	BW										
2. Check tightness and adjustment of thrust collars on sleeve bearing units and locking collars on ball bearing units.		RA	BW										
3. Check belt tension and adjust if necessary.		RA	BW										
4. Clean strainer (if atmosphere is extremely dirty, it may be necessary to clean strainer weekly).		RA	BW										
5. Check for biological growth in sump. Consult water treatment specialist if such growth is not under control.		RA	BW										
6. Clean and flush sump.		RA	BW										
7. Check spray distribution system. Check and reorient nozzles, if necessary. On evaporative condensers and industrial fluid coolers with trough type distribution systems, adjust and flush out troughs if necessary.		RA	BW										
8. Check operating water level in the pan and adjust float valve if required.		RA	BW										
9. Check bleed-off rate and adjust if necessary.		RA	BW										
10. Check fans and air inlet screens and remove any dirt or debris.		RA	BW										
ONCE EVERY YEAR	Inspect and clean protective finish inside and out. Look particularly for any signs of spot corrosion. Clean and refinish any damaged protective coating.												
Before undertaking start-up procedures or performing inspection or maintenance of equipment, make certain the power has been disconnected. Refer to appropriate operating and maintenance manuals and comply with all caution label instructions.													

______________ **1.** The preventive maintenance checklist procedure is for a(n) ___.

______________ **2.** Where would the equipment be commonly located?

_______________ **3.** What type of mechanical cooling equipment uses this device?

_______________ **4.** When was preventive maintenance last performed?

_______________ **5.** If it is currently June, how many months have been skipped?

_______________ **6.** If this procedure takes 2 hr per month at a billed rate of $75 per hour, how much money should be budgeted per year for the preventive maintenance labor on this unit?

Section 4.1 Indoor Air Quality

REVIEW QUESTIONS

Name ______________________ **Date** ______________

Multiple Choice

______ **1.** The ___ ranks IAQ as one of the top five environmental threats to human health.
A. Centers for Disease Control (CDC)
B. Environmental Protection Agency (EPA)
C. Occupational Safety and Health Administration (OSHA)
D. National Fire Protection Association (NFPA)

______ **2.** ___ is the process of introducing fresh air into a building.
A. Ventilation
B. Compression
C. Filtration
D. Enthalpy

______ **3.** A common percentage of ___ air used for ventilation are 5%, 10%, and 30%.
A. mixed
B. return
C. inside
D. outside

______ **4.** ___ air is air that contains odors or contaminants.
A. Return
B. Outside
C. Stale indoor
D. Mixed

______ **5.** Reducing the amount of outside air brought into a building means an increase in recirculated indoor air and increase in ___.
A. financial costs
B. energy consumption
C. indoor air quality
D. adverse health effects

______ **6.** The increased use of outside air for ventilation may lead to ___ for a building.
A. increased energy costs
B. reduced IAQ problems
C. reduced risk of freezing
D. increased CO_2 levels

_______________ **7.** ___ is the most common issue cited for indoor air pollution injuries.
- A. Negligence
- B. Construction eviction
- C. Workers' compensation
- D. Termination of landlord/tenant relationship

_______________ **8.** The American National Standards Institute (ANSI) and American Society of Heating, Refrigerating, and Air-Conditioning Engineers (ASHRAE) have enacted standards for ___ in commercial buildings.
- A. humidity
- B. room temperature
- C. IAQ
- D. airflow velocity

_______________ **9.** ___ is a designation of the contaminants present in the air.
- A. Ventilation
- B. Indoor air quality (IAQ)
- C. Ambient comfort
- D. Electrostatic filtration

_______________ **10.** The amount of outside air admitted, measured in ___, is regulated by ASHRAE Standard 62.1-2010.
- A. meters per second (mps) of air
- B. cubic feet per minute per person (cfm/person) of air
- C. percent (%) of contamination
- D. parts per million (ppm) of CO_2

Completion

_______________ **1.** ___ is a designation of contaminant levels present in the air inside buildings.

_______________ **2.** ___ indoor air is air that contains odors or contaminants.

_______________ **3.** ___ is the minimum amount of outside air that must be mixed with return air before the air is allowed to enter a building space.

_______________ **4.** Good ___ means that the air is free of harmful particles or chemicals.

_______________ **5.** ___ is the most common issue cited for indoor air pollution injuries.

_______________ **6.** ___ air is the combination of return air and outside air.

Section 4.2 Air Handling Units

REVIEW QUESTIONS

Name ______________________________ Date ____________________

Multiple Choice

______________ **1.** A(n) ___ is a device consisting of a fan, ductwork, filters, dampers, heating coils, cooling coils, humidifiers, sensors, and controls that conditions and distributes air throughout a building.

A. exhaust fan
B. controller
C. air handling unit
D. variable air volume terminal box

______________ **2.** A(n) ___ air system is a system that uses a specified amount of combined return air and outside air to condition building spaces.

A. economizer
B. inside
C. mixed
D. 100% outside

______________ **3.** The most common fan used in commercial building air handling units is a(n) ___ fan.

A. exhaust
B. centrifugal
C. return
D. reciprocating

______________ **4.** A ___ is a porous material that removes particles from a moving fluid.

A. filter
B. vent
C. duct
D. damper

______________ **5.** A ___ is a device that adds moisture to the air by causing water to evaporate into the air.

A. filter
B. humidifier
C. heating coil
D. cooling coil

______________ **6.** A ___ air handling unit is an air handling unit that moves a constant volume of air.

A. constant-volume
B. multizone
C. dual-duct
D. variable air volume

_______________ **7.** A(n) ___ air handling unit is an air handling unit that moves a variable volume of air.
A. reciprocating
B. induction
C. scroll
D. variable air volume

_______________ **8.** The most important function of a variable air volume air handling unit is the ability to change the volume of air produced by the ___.
A. supply fan
B. exhaust fan
C. vortex damper
D. cooling coils

_______________ **9.** A ___ is a device that controls the airflow to a building, matching the building space requirements for comfort.
A. constant-volume air handling unit
B. mixed air handling unit
C. variable air volume (VAV) terminal box
D. constant-speed blower

_______________ **10.** ___ are used to ensure airflow across the heating element during the heating mode.
A. Electric motor variable-frequency drives
B. Centrifugal fans
C. Fan-powered VAV terminal boxes
D. Vortex dampers

_______________ **11.** A hybrid air handling unit is an air handling unit that is ___.
A. part induction and part terminal reheat
B. part multizone and part dual-duct
C. an induction air handling unit with a mixing box
D. a multizone unit with filters for each zone

_______________ **12.** A multizone air handling unit has ___ for each zone that mix the hot and cold air and send it through separate ducts to the appropriate zones.
A. dampers
B. flow control valves
C. small fans
D. filters

_______________ **13.** A(n) ___ damper is a pie-shaped damper located at the inlet of a centrifugal fan.
A. exhaust air
B. return air
C. outside air
D. vortex

__________ 14. Air handling units can be categorized into ___.
A. primary loop air systems and secondary loop systems
B. 100% outside air systems and mixed air systems
C. mechanical draft systems and natural draft systems
D. return air systems and exhaust air systems

Completion

__________ 1. A(n) ___ system is a system that does not recirculate any return air from building spaces.

__________ 2. A(n) ___ is a device with rotating blades or vanes that move air.

__________ 3. ___ is the distribution system for a forced air heating or cooling system.

__________ 4. A(n) ___ is a device that cleans air by passing the air through electrically charged plates and collector cells.

__________ 5. A(n) ___ is an adjustable metal blade or set of blades used to control the flow of air.

__________ 6. A(n) ___ is a finned heat exchanger that adds heat to the air flowing over it.

__________ 7. A(n) ___ cools and dehumidifies incoming fresh air in the summer and heats and humidifies incoming air in the winter.

__________ 8. A(n) ___ air handling unit is an air handling unit that provides heating, ventilation, and air conditioning to only one building zone or area.

__________ 9. A(n) ___ air handling unit is an air handling unit that has hot and cold air ducts connected to mixing boxes at each building space.

__________ 10. A(n) ___ air handling unit is an air handling unit that delivers air at a constant 55°F temperature to building spaces.

__________ 11. A(n) ___ air handling unit is an air handling unit that maintains a constant 55°F air temperature and delivers the air to the building spaces at a high duct pressure.

__________ 12. A(n) ___ is an electronic device that controls the direction, speed, and torque of an electric motor.

__________ 13. ___ contain components that move, filter, control airflow direction, heat, cool, humidify, and dehumidify the air in building spaces.

__________ 14. A(n) ___ air handling unit is an air handling unit that is designed to provide heating, ventilation, and air conditioning for more than one building zone or area.

Air Handling Unit

______________________ **1.** mixed air plenum

______________________ **2.** return air

______________________ **3.** humidifier

______________________ **4.** damper actuators

______________________ **5.** ductwork

______________________ **6.** outside air

______________________ **7.** return air damper (NO)

______________________ **8.** exhaust air

______________________ **9.** drain

______________________ **10.** mixed air

______________________ **11.** outside air damper (NC)

______________________ **12.** exhaust air damper (NC)

______________________ **13.** supply fan

______________________ **14.** temperature controller in building space

______________________ **15.** steam supply

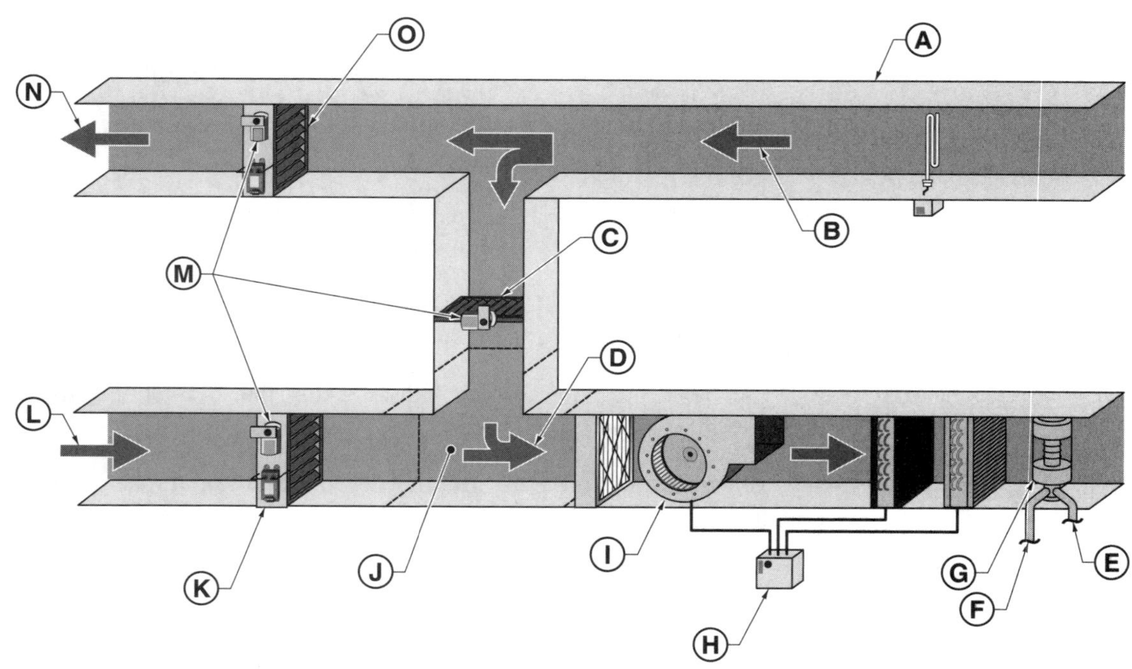

ACTIVITIES

Name ______________________________ **Date** ____________________

Activity 4-1. Commercial HVAC Systems

Your firm has taken over the operations and maintenance contract on a commercial facility. In preparation for the beginning of the contract, the type of HVAC equipment must be determined and basic questions answered. Upon arriving at the facility, it is discovered that there are four air handling units in the fan room. Each has an identical diagram. Use the air handling unit drawing to answer the questions.

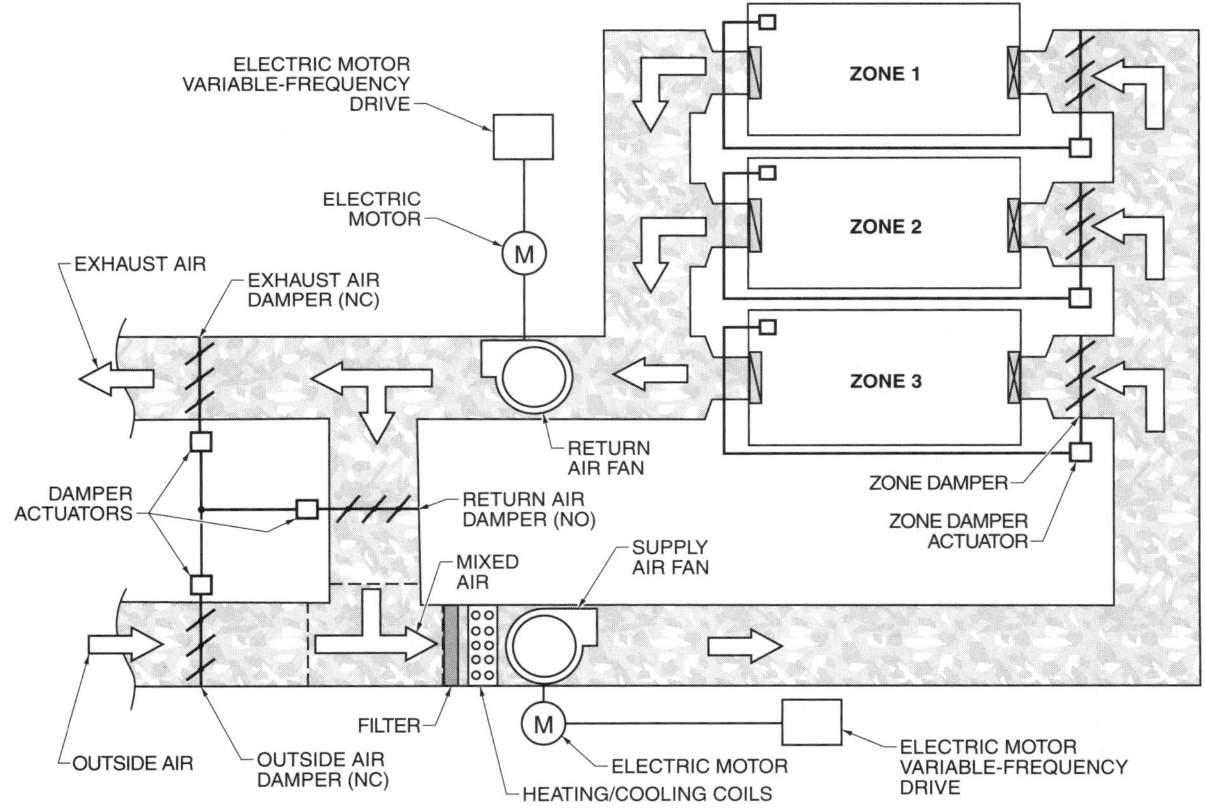

__________________ **1.** What type of air handling unit is it?

__________________ **2.** Is the air handling unit a 100% outside air system or a mixed air system?

______________________ **3.** Is economizer cooling available with the unit?

______________________ **4.** If indoor air quality is a concern, what may be introduced to reduce IAQ problems?

______________________ **5.** What is a common duct static pressure?

______________________ **6.** The unit provides air for ___ (number) zones.

______________________ **7.** If each air handling unit is identical, what is the total number of zones?

______________________ **8.** If a trap is not present on the heating coil, what type of heat is being used?

______________________ **9.** What device removes particulate matter from the air stream?

______________________ **10.** Preventive maintenance has been estimated at ½ hr per month per terminal box and 1 hr for each air handling unit. How many total hours of preventive maintenance per month are required for all of the air handling units and zone terminal boxes?

Section 5.1 Heating System Energy Sources

REVIEW QUESTIONS

Name ______________________________ **Date** ____________________

Multiple Choice

______________ **1.** In most HVAC applications, it is more expensive to heat a building space with ___ than with other energy sources.

A. electricity
B. natural gas
C. fuel oil
D. solar energy

______________ **2.** ___ is the intensity of heat required to start a chemical reaction.

A. Btu
B. Emergency heat
C. Viscosity
D. Ignition temperature

______________ **3.** ___ is the number of British thermal units (Btu) per pound or gallon of fuel.

A. Temperature
B. Heating value
C. CFC
D. Fuel oil number

______________ **4.** The amount of energy received at the surface of the Earth can exceed ___ per sq ft of surface, depending on the angle of the sun's rays and the position of the solar collector.

A. 100 Btu/hr
B. 200 Btu/hr
C. 300 Btu/hr
D. 400 Btu/hr

______________ **5.** Commercial building heating system failure results in ___.

A. melted pipes
B. the rusting of pipes
C. a loss of electricity in the building
D. occupant discomfort

______________ **6.** Fuel, oxygen, and heat are required for ___.

A. emergency heat
B. heating
C. solar energy
D. combustion

_______________ **7.** The ___ has established standards for grading fuel oil.
A. National Fire Protection Association (NFPA)
B. American Society of Heating, Refrigeration and Air-Conditioning Engineers (ASHRAE)
C. American Society for Testing and Materials (ASTM)
D. American National Standards Institute (ANSI)

_______________ **8.** ___ are used in electric baseboard heaters, radiant heat panels, air handling units (AHUs), and variable air volume (VAV) terminal boxes.
A. Absorbers
B. Compressors
C. Electric heating elements
D. Heat pumps

_______________ **9.** The purpose of an HVAC system is to provide ___ to the occupants of a building space.
A. electricity
B. warm air
C. comfort
D. clean air

_______________ **10.** ___ can be designed for easy movement to new locations in a building.
A. Air handling units
B. Radiant heat panels
C. Electric baseboard heaters
D. Variable-air-volume (VAV) terminal boxes

_______________ **11.** ___ is an odorless and colorless vapor at standard atmospheric temperature conditions.
A. Propane
B. Natural gas
C. Fuel oil
D. Sulfur dioxide

_______________ **12.** Fuel oil tank heaters and line heaters must be used to heat No. ___ fuel oil to the required temperature for transport and combustion.
A. 2
B. 6
C. 7
D. 9

_______________ **13.** The reversal of the flow of refrigerant enables a heat pump to transfer heat from the outdoors to the indoors to produce a ___.
A. flow imbalance
B. backflow
C. cooling effect
D. heating effect

_______________ **14.** A(n) ___ is a device that consists of wire coils that become hot when energized.
A. expansion device
B. spool
C. electric heating element
D. fuel oil pump

_______________ **15.** Emergency heat is normally provided by ___.
A. electric heating elements
B. heat pumps
C. gas-fired boilers
D. air handling units

Completion

_______________ **1.** ___ is commonly used as an energy source for heating commercial buildings because it is plentiful and relatively inexpensive.

_______________ **2.** ___ is the chemical reaction that occurs when oxygen reacts with the hydrogen (H) and carbon (C) present in a fuel at ignition temperature.

_______________ **3.** ___ is the ability of a liquid to resist flow.

_______________ **4.** The ___ required depends on the type of burner and whether a straight distillate fuel oil or a blend of fuel oils is used.

_______________ **5.** ___ is the heat created by the visible (light) and invisible (infrared) energy rays of the sun.

_______________ **6.** ___ is heat provided if the outside air temperature drops below a set temperature or if a heat pump fails.

_______________ **7.** The four grades of fuel oil used in boilers are No. ___ fuel oil, No. 4 fuel oil, No. 5 fuel oil, and No. 6 fuel oil.

_______________ **8.** A(n) ___ is a device that consists of wire coils that become hot when energized.

Fuel Oil System

______ 1. fuel oil relief valve

______ 2. fuel oil from tank

______ 3. fuel oil strainer

______ 4. main fuel oil solenoid valve

______ 5. fuel oil burner pressure gauge

______ 6. nozzle air pressure gauge

______ 7. fuel oil pump

______ 8. burner nozzle

______ 9. atomizing air

______ 10. fuel oil thermometer

______ 11. fuel oil pressure gauge

______ 12. fuel oil pressure regulator

______ 13. check valve

______ 14. fuel oil returned to tank

______ 15. shutoff valve

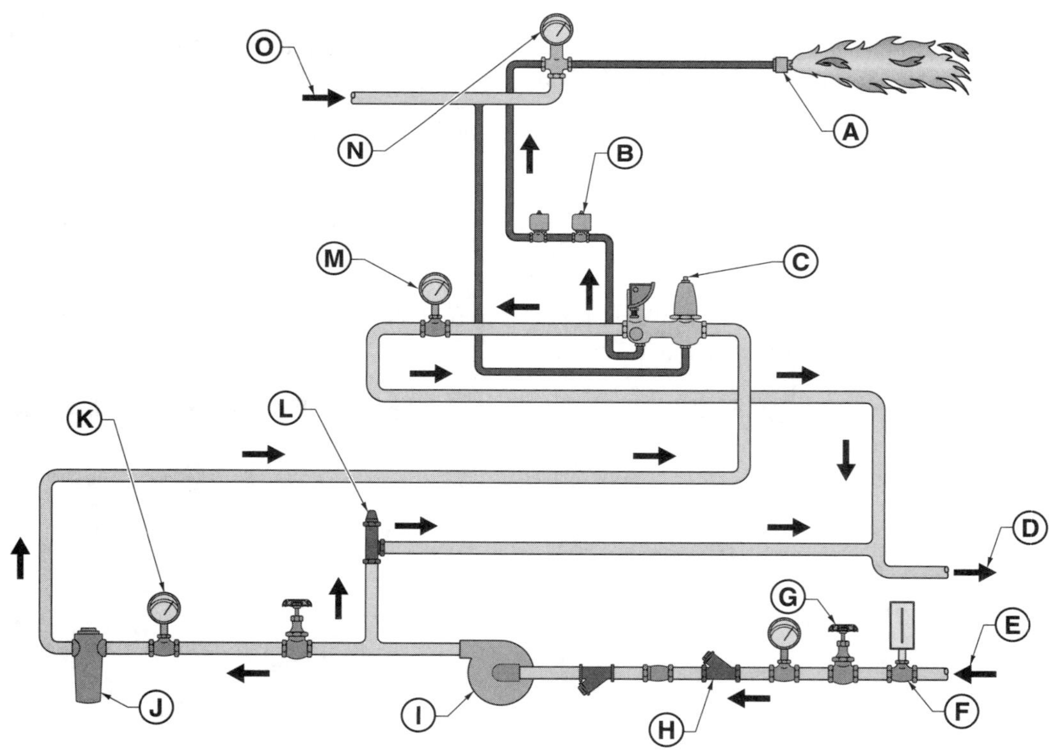

Section 5.2 Cooling System Energy Sources

REVIEW QUESTIONS

Name ______________________________ **Date** ______________________

Multiple Choice

________________ **1.** An air conditioning system is a system that produces a(n) ___ effect and distributes the cool air or water to building spaces.
- A. refrigeration
- B. heating
- C. open flow
- D. heat pump

________________ **2.** A refrigerant and ___ are required in absorption refrigeration systems.
- A. expansion device
- B. compressor
- C. generator
- D. absorbent

________________ **3.** Steam or ___ is used to provide cooling in absorption refrigeration systems.
- A. hot water
- B. cold water
- C. evaporation
- D. heat

________________ **4.** The use of outside air for cooling is referred to as ___ cooling.
- A. free
- B. low-cost
- C. energy-efficient
- D. ambient

________________ **5.** Absorption refrigeration systems have a(n) ___ in place of the compressor to raise system pressure.
- A. heat exchanger
- B. expansion device
- C. evaporator and condenser
- D. generator and absorber

________________ **6.** The basic energy source used to cool a building is ___ air.
- A. mixed
- B. return
- C. outside
- D. circulated

Completion

______________ **1.** A(n) ___ is a mechanical device that compresses refrigerant or other fluid.

______________ **2.** A(n) ___ is a system that uses a liquid (normally water) to cool building spaces.

______________ **3.** A(n) ___ is a refrigeration system that uses the absorption of one chemical by another chemical and heat transfer to produce a refrigeration effect.

______________ **4.** Steam and hot water cooling systems normally do not use ___, avoiding the problem of refrigerant replacement.

______________ **5.** The mechanical equipment of a(n) ___ system can be used to transfer heat from the air inside a building to the air outside a building, producing a cooling effect.

______________ **6.** ___ is the process of cooling the air in building spaces to provide a comfortable temperature.

______________ **7.** ___ compression systems are commonly used for large commercial or industrial applications where mechanical refrigeration systems are not as efficient.

______________ **8.** The applications of alternative energy sources depend largely on local availability of fuel, construction codes, and ___ standards.

Refrigeration System

______________ **1.** condenser

______________ **2.** expansion device

______________ **3.** suction line

______________ **4.** evaporator

______________ **5.** hot gas discharge line

______________ **6.** compressor/motor

______________ **7.** liquid line

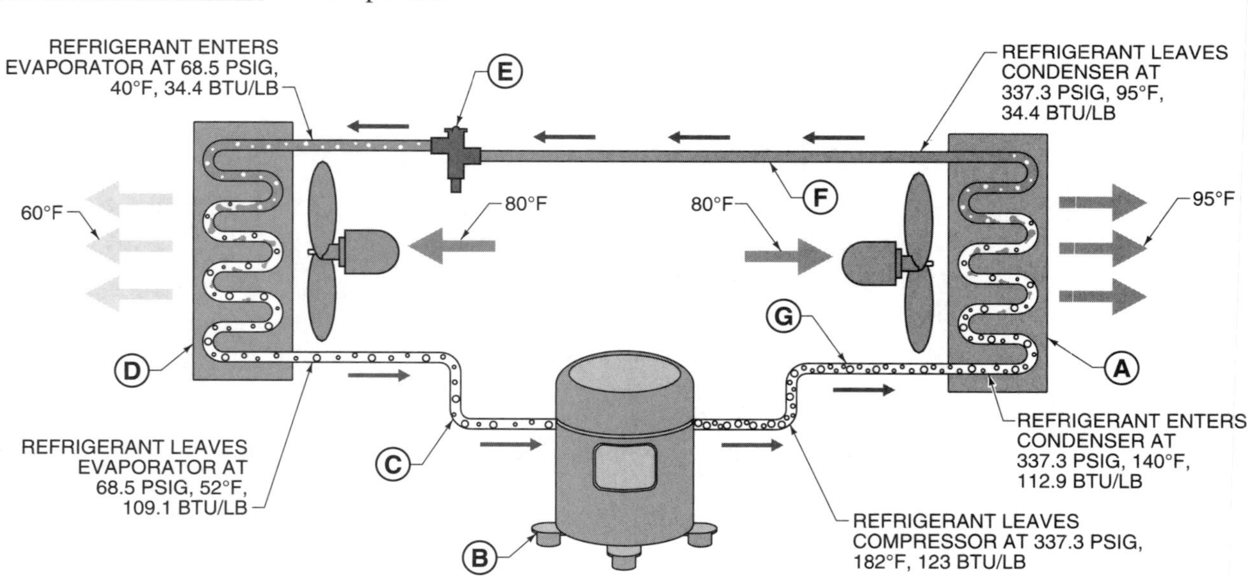

Liquid Chiller

______ 1. cooling tower

______ 2. warm water out (95°F)

______ 3. liquid chiller

______ 4. expansion device

______ 5. louvers

______ 6. air outlet

______ 7. cool water collection

______ 8. compressor

______ 9. hot water inlet

______ 10. condenser (tube-in-shell) heat exchanger

______ 11. condenser water pump

______ 12. cool water in (85°F)

______ 13. evaporator (tube-in-shell) heat exchanger

______ 14. refrigerant flow

______ 15. air inlet

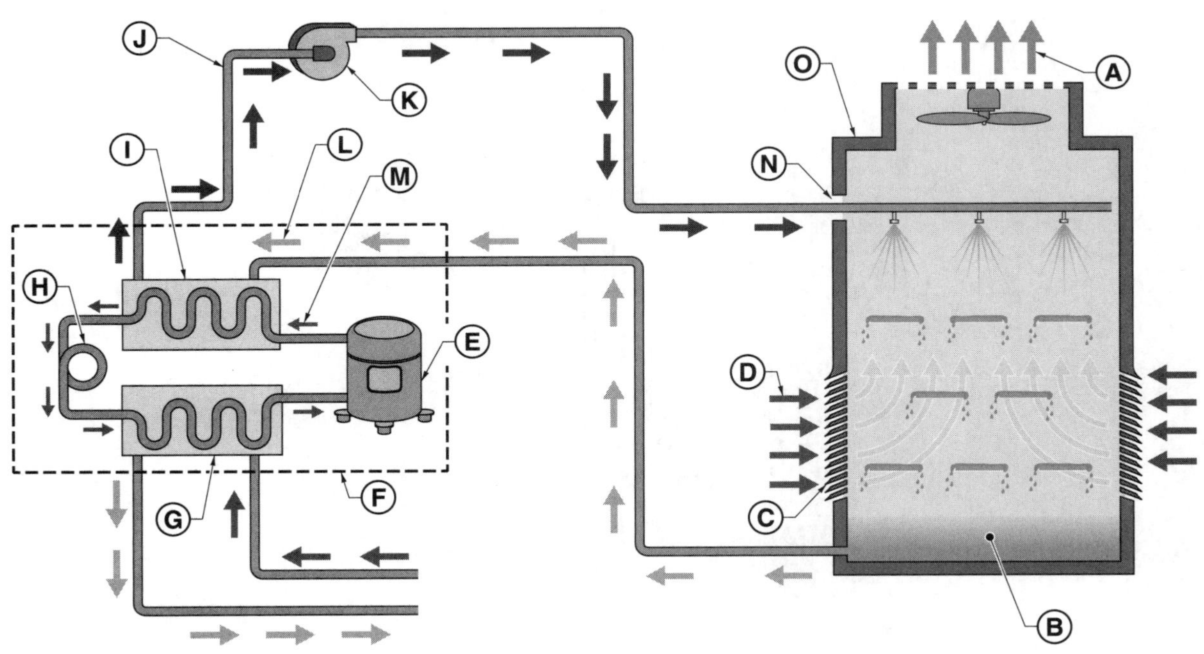

Heat Pump Cooling Cycle

_________________ **1.** hot outside air

_________________ **2.** cool air to building space

_________________ **3.** reversing valve

_________________ **4.** compressor

_________________ **5.** warm air from building space

_________________ **6.** outdoor unit

_________________ **7.** spool

_________________ **8.** expansion device

_________________ **9.** warm outside air

_________________ **10.** indoor unit

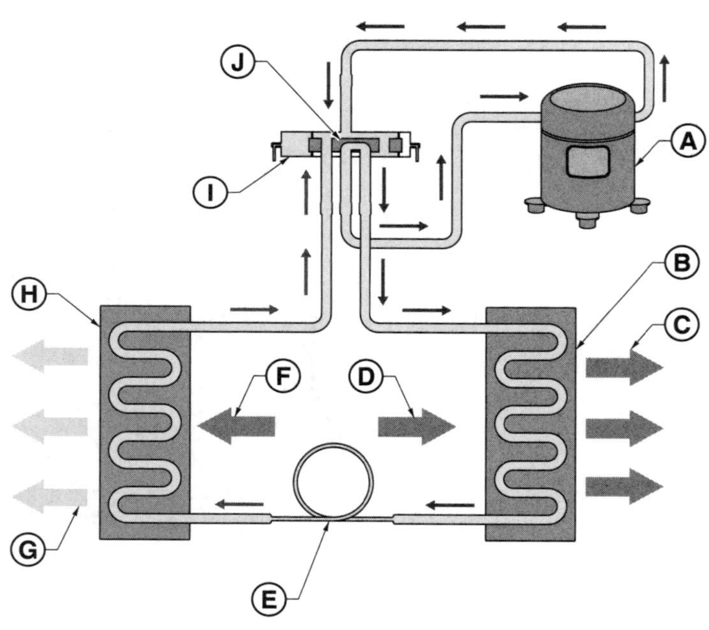

CHAPTER 5

HVAC System Energy Sources

ACTIVITIES

Name ______________________________ Date ____________________

Activity 5-1. Absorption Cooling Selection

Absorption chiller information must be determined from the manufacturer's catalog. The absorber model number is T1B-ST-9E2-28-A-S. Use the absorption chiller nomenclature to answer the questions. Note: *More data would be included with the absorber.*

Model Nomenclature

Note: The model number denotes the following characteristics of the unit:

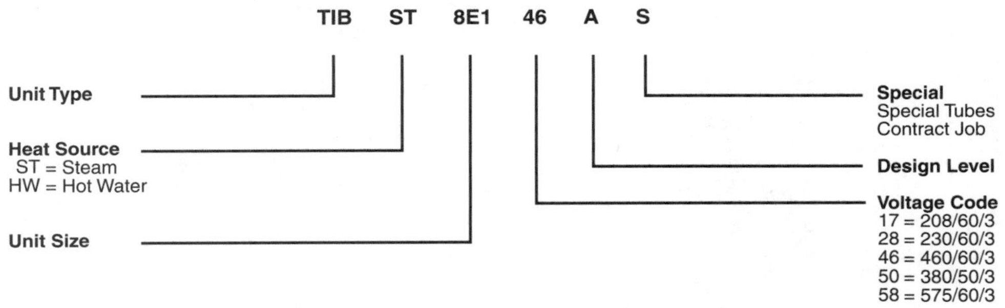
TIB ST 8E1 46 A S
Unit Type
Heat Source
ST = Steam
HW = Hot Water
Unit Size
Special
Special Tubes
Contract Job
Design Level
Voltage Code
17 = 208/60/3
28 = 230/60/3
46 = 460/60/3
50 = 380/50/3
58 = 575/60/3

______________ **1.** The manufacturer unit type designation for the absorber is ___.

______________ **2.** The heat source is ___.

______________ **3.** The unit size is ___.

______________ **4.** The voltage code used is ___.

______________ **5.** The design level designation is ___.

______________ **6.** Does the absorber have special tubes, or is it a contract job?

Activity 5-2. Heat Pump Selection

Work is required in a new wing of a building that contains heat pumps. The first heat pump nameplate is listed as W-CDD-2-015-F-Z. Use the model nomenclature to answer the questions.

Model Nomenclature

Note: For illustration purposes only. Not all options available with all models.
Please consult Water Source Heat Pump Representative for specific availability.

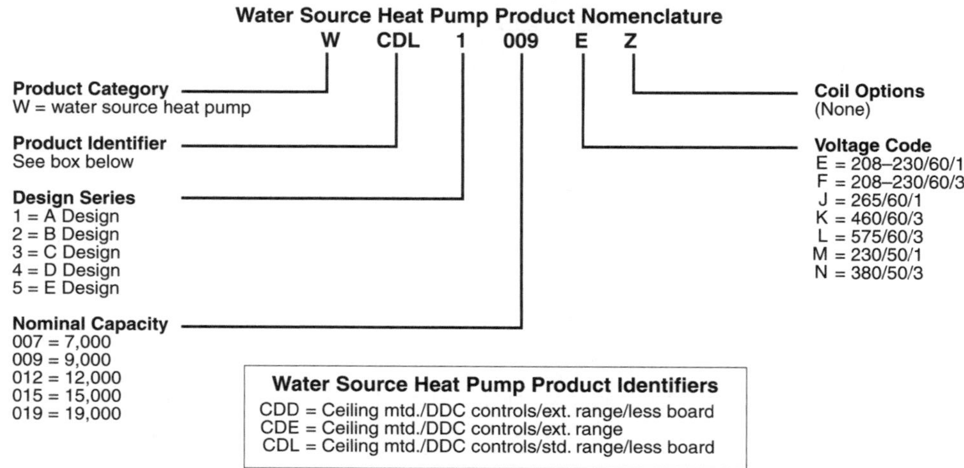

MOTOMASTER CORPORATION

PE•21 PLUS™		PREMIUM EFFICIENCY	
ID. NO.	P18S3030	MOTOR WEIGHT	70.5 LBS
TYPE	RGZESD I	FRAME	182T
HP	3.00	SERVICE FACTOR	1.15 3 PH
AMPS	7.9/3.95	VOLTS	230/460
RPM	1750	HERTZ	60
DUTY	CONT 40°C AMB.	ENCL.	IEFC
CLASS INSUL	F NEMA DESIGN B TYPE P K.V.A. CODE K	NEMA NOM. EFF.	87.5 GUARANT'D. EFFICIENCY 85.5
DRIVE END BEARING	30BC02XPP30X26	OPP. END BEARING	25BC02XPP30X26
MAX. CORR. kVAR	1.0	POWER FACTOR	82.0

51-770-642

LOW VOLTS: T4 T5 T6 / T7 T8 T9 / T1 T2 T3 / L1 L2 L3

HIGH VOLTS: T4 T5 T6 / T7 T8 T9 / T1 T2 T3 / L1 L2 L3

Yale, Illinois 60070-4300

made in U.S.A.

______________ **1.** Is the heat pump a water source heat pump?

______________ **2.** Does the heat pump use DDC controls?

______________ **3.** The heat pump nominal capacity is ___ Btu/hr.

______________ **4.** If there are 12,000 Btu/hr/ton, how many tons of cooling can be achieved by the heat pump?

______________ **5.** If the voltage applied is 208 VAC–230 VAC, 60 Hz, 1ϕ, does this agree with the listing on the nameplate?

Section 6.1 Control System Components

REVIEW QUESTIONS

Name ______________________________ **Date** ____________________

Multiple Choice

______________ **1.** A(n) ___ is a device that measures a controlled variable such as temperature, pressure, or humidity and sends a signal to a controller.

A. controlled device
B. anticipator
C. actuator (valve or damper)
D. sensor

______________ **2.** Toxins and pollutants may build up to unacceptable levels, causing health problems to occupants, if enough ___ is not circulated in a building space.

A. aerosol
B. desiccant
C. fresh air
D. return air

______________ **3.** The most common ___ are hot water, chilled water, steam, hot air, and cold air.

A. setpoints
B. control agents
C. transmitters
D. all of the above

______________ **4.** A ___ is a device that receives a signal from a sensor, compares it to a setpoint value, and sends an appropriate output signal to a controlled device.

A. controller
B. hub
C. router
D. control agent

Completion

______________ **1.** A(n) ___ is an arrangement of a sensor, controller, and controlled device to maintain a specific controlled variable value in a building space, pipe, or duct.

______________ **2.** A(n) ___ is the object that regulates the flow of fluid in a system to provide the heating, air conditioning, or ventilation effect.

______________ **3.** A(n) ___ is fluid that flows through controlled devices to produce a heating or cooling effect in the system or building spaces.

______________ **4.** ___ controls are always used with humidification systems to prevent excessively high humidity levels in a duct or building space.

______________ **5.** Cooling systems use chilled water, direct expansion cooling, or outside air as ___.

______________ **6.** Common ___ include dampers for regulating airflow, valves for regulating water or steam flow, refrigeration compressors for delivering cooling, and gas valves and electric heating elements for regulating heating.

Economizer Cooling

______________ **1.** filter

______________ **2.** exhaust air dampers open

______________ **3.** return air dampers closed

______________ **4.** supply fan

______________ **5.** outside air dampers open

______________ **6.** chilled water valve closed

______________ **7.** cooling coil

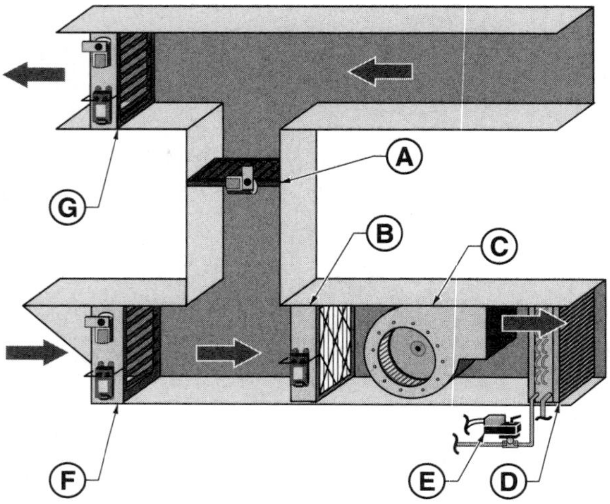

A
G
B
C
F
E
D

Dehumidification

______________ **1.** chilled water valve

______________ **2.** cool, dry air

______________ **3.** cooling coil

______________ **4.** drain pan and drain line to remove condensate

______________ **5.** hot, humid air

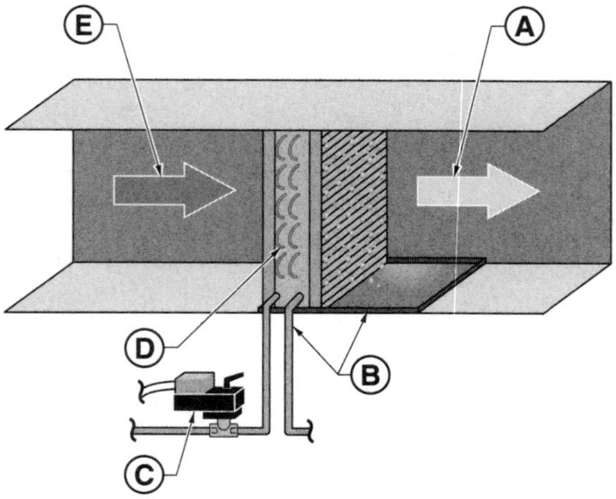

E
A
D
B
C

Section 6.2 Control Systems

REVIEW QUESTIONS

Name ______________________________ **Date** ____________________

Multiple Choice

______________ **1.** ___ is the desired value to be maintained by a system.
A. Control point
B. Setpoint
C. Offset
D. Working point

______________ **2.** A(n) ___ is the actual value that a control system experiences at any given time.
A. control point
B. setpoint
C. offset
D. controlled variable

______________ **3.** ___ is control in which a controller produces only a 0% or 100% output signal.
A. Closed loop control
B. Anticipator control
C. ON/OFF (digital) control
D. Overshooting control

______________ **4.** HVAC electrical controls often use ___ VAC for proper operation of the control system.
A. 6
B. 24
C. 115
D. 230

______________ **5.** ___ is the decreasing of a controlled variable below the controller setpoint.
A. Undershooting
B. Overshooting
C. Offsetting
D. Undercutting

______________ **6.** A variable-speed drive permits volume control of the ___.
A. humidifier
B. return fan
C. heating coil
D. supply fan or pump

_______________ **7.** In system pressure control, the ___ is connected to a controller which opens or closes a controlled device such as a damper or valve.
A. control point
B. humidity control
C. pressure sensor
D. differential pressure control

_______________ **8.** In differential pressure control, a specific pressure in the system is maintained based on a ___.
A. specific setpoint
B. difference in pressure between two points in the system
C. the offset between the setpoint and actual air temperature
D. pressure control valve

_______________ **9.** The most common method of ___ control is to include a humidity sensor or controller in the building space.
A. building space temperature
B. return air
C. humidity
D. volume

Completion

_______________ **1.** ___ temperature control is more accurate than building space temperature control because the return air is at a temperature which is an average of all the temperatures in the building space.

_______________ **2.** ___ is the difference between a control point and a setpoint.

_______________ **3.** ___ is the measurement of the results of a controller action by a sensor or switch.

_______________ **4.** A(n) ___ is a device that turns heating or cooling equipment ON or OFF before it normally would.

_______________ **5.** ___ is control in which the controlled device is positioned in direct response to the amount of offset in the system.

_______________ **6.** Once adjusted, a control system should operate ___.

_______________ **7.** ___ control is one of the most important components of comfort.

_______________ **8.** ___ controllers include air conditioning compressors, electric heating stages, gas valves, refrigeration compressors, and constant-speed fans.

_______________ **9.** At any given time, the ___ may be different from the control point.

_______________ **10.** ___ is the increasing of a controlled variable above the controller setpoint.

_______________ **11.** ___ loop control is control in which no feedback occurs between the controller, sensor, and controlled device.

_______________ 12. ___ loop control is control in which feedback occurs between the controller, sensor, and controlled device.

Volume Control

_______________ 1. linkage

_______________ 2. fan housing

_______________ 3. actuator

_______________ 4. inlet vane

_______________ 5. inlet vane crank arm

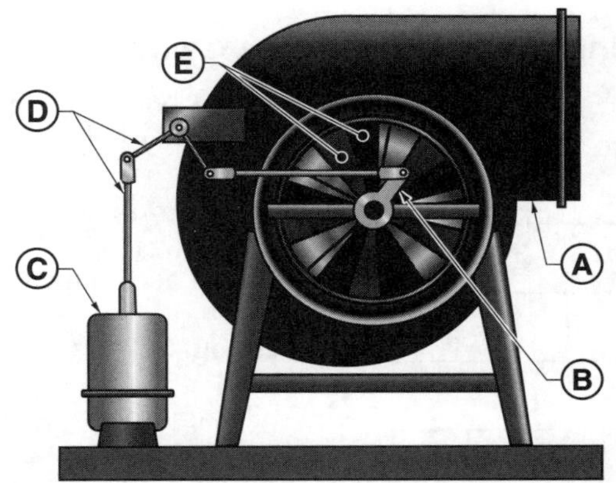

Open Loop Control

_______________ 1. chilled water pump

_______________ 2. chiller evaporator (heat exchanger)

_______________ 3. pump contactor

_______________ 4. outside air thermostat set at 65°F

_______________ 5. chilled water supply

_______________ 6. chiller compressor

_______________ 7. chilled water return

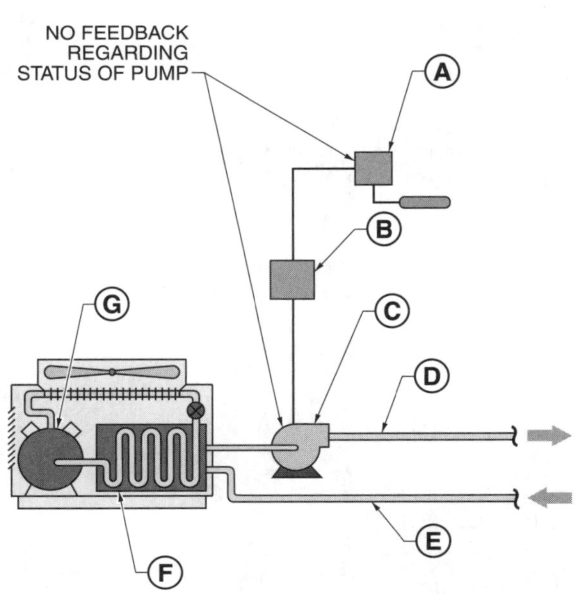

Offset

_______________ **1.** control point

_______________ **2.** setpoint

_______________ **3.** offset

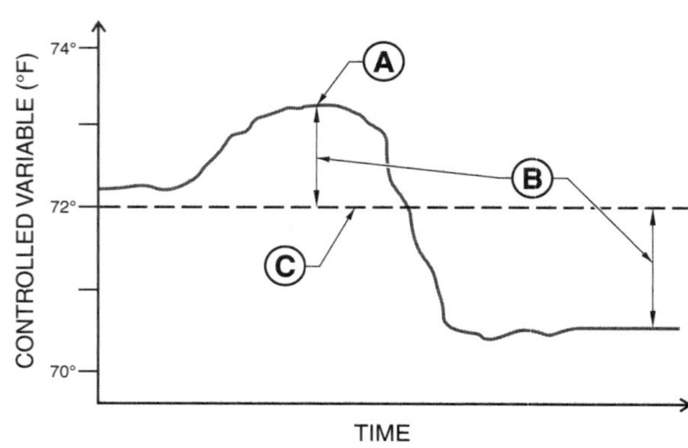

Building Space Temperature Control

_______________ **1.** filter

_______________ **2.** supply air duct

_______________ **3.** exhaust air damper

_______________ **4.** heating coil

_______________ **5.** building space

_______________ **6.** return air duct

_______________ **7.** outside air damper

_______________ **8.** humidifier

_______________ **9.** return fan

_______________ **10.** cooling coil

_______________ **11.** room controller or thermostat

_______________ **12.** supply fan

_______________ **13.** return air damper

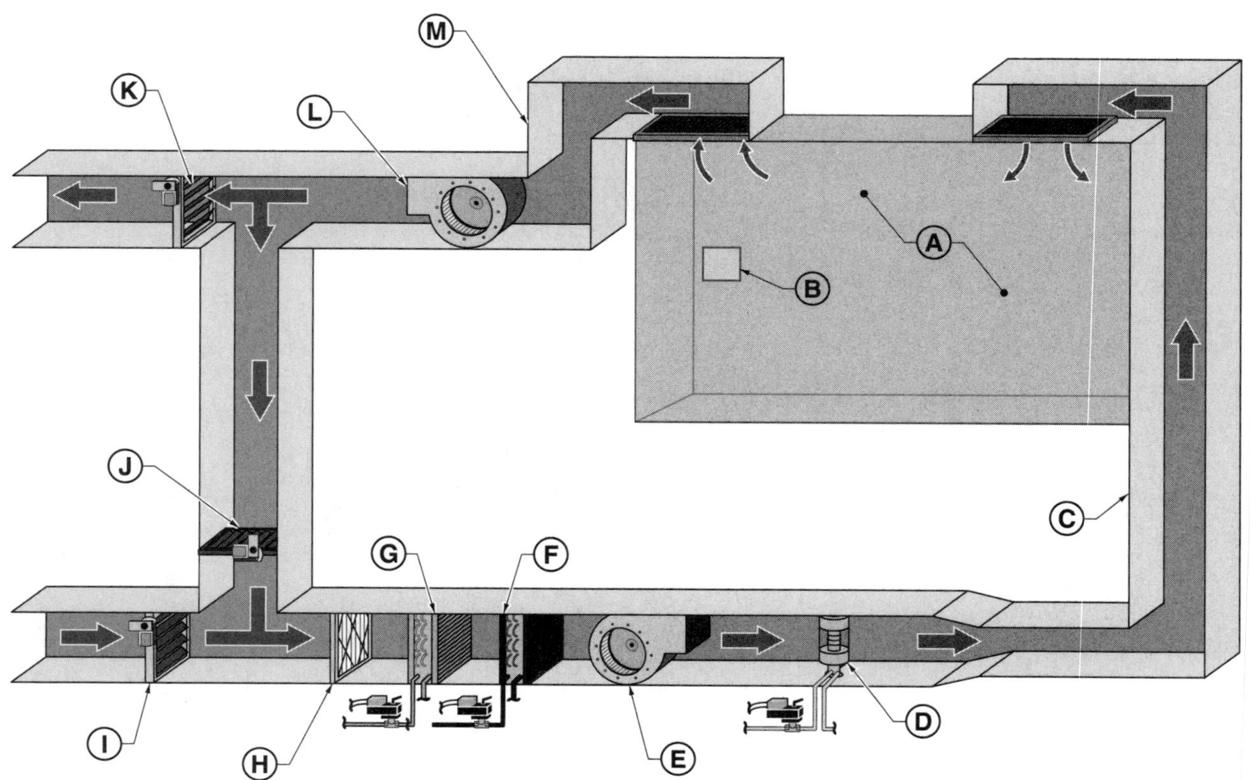

Control Principles

ACTIVITIES

Name ______________________________ **Date** ________________

Activity 6-1. Control Principles

The control system of a new building has not been wired.

1. Complete the drawing of the room temperature control system. Add wiring to connect the components to make the control system function.

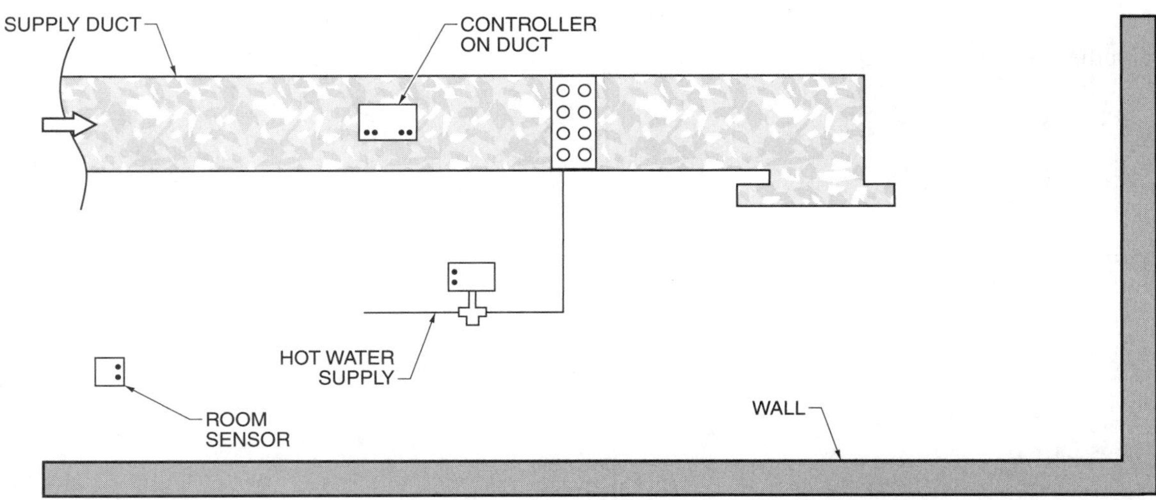

________________ **2.** Does the drawing show open or closed loop control?

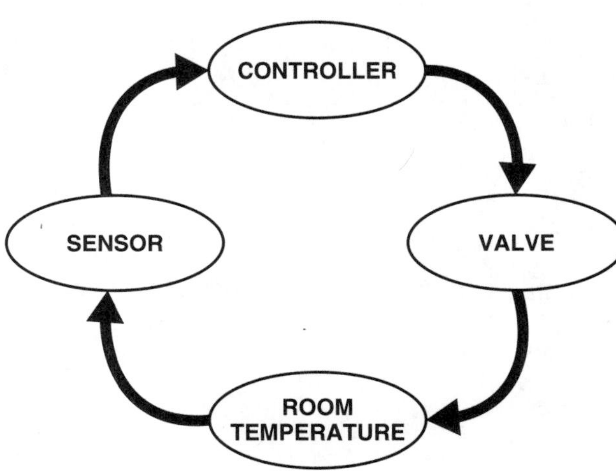

After connecting the devices, the HVAC unit begins to operate. After a few hours, the occupants complain that the temperature is erratic. The room temperature is checked every 5 min to determine system operation. The room temperature is checked with an accurate electronic temperature sensor.

3. Enter the offset value for each temperature measurement.

TEMPERATURE READINGS			
TIME	TEMPERATURE*	SETPOINT*	OFFSET*
8:15 AM	78	74	
8:20 AM	75	74	
8:25 AM	72	74	
8:30 AM	70	74	
8:35 AM	68	74	
8:40 AM	71	74	

* in °F

4. Graph the offset on the chart.

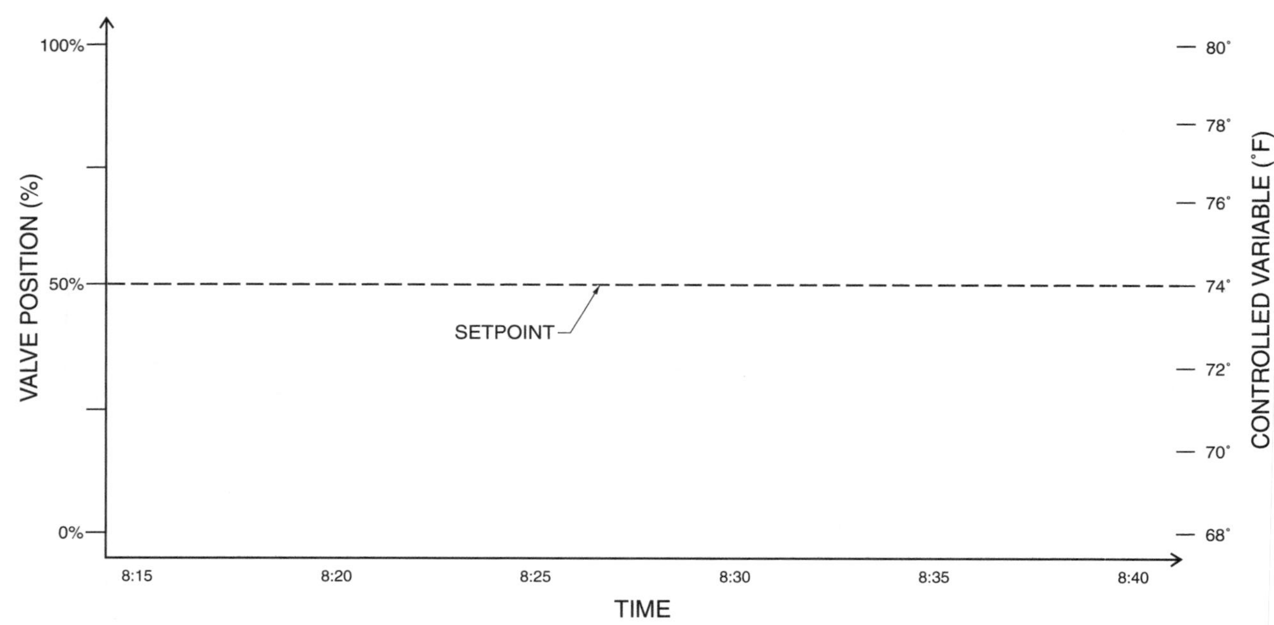

________________ **5.** Is this level of control acceptable?

After making some controller adjustments, the control appears better. The room temperature is checked again every 5 min to determine system operation.

6. Enter the offset value for each temperature measurement.

TEMPERATURE READINGS			
TIME	TEMPERATURE*	SETPOINT*	OFFSET*
9:15 AM	75	74	
9:20 AM	74.5	74	
9:25 AM	74	74	
9:30 AM	75	74	
9:35 AM	75.5	74	
9:40 AM	75	74	

* in °F

7. Graph the offset on the chart.

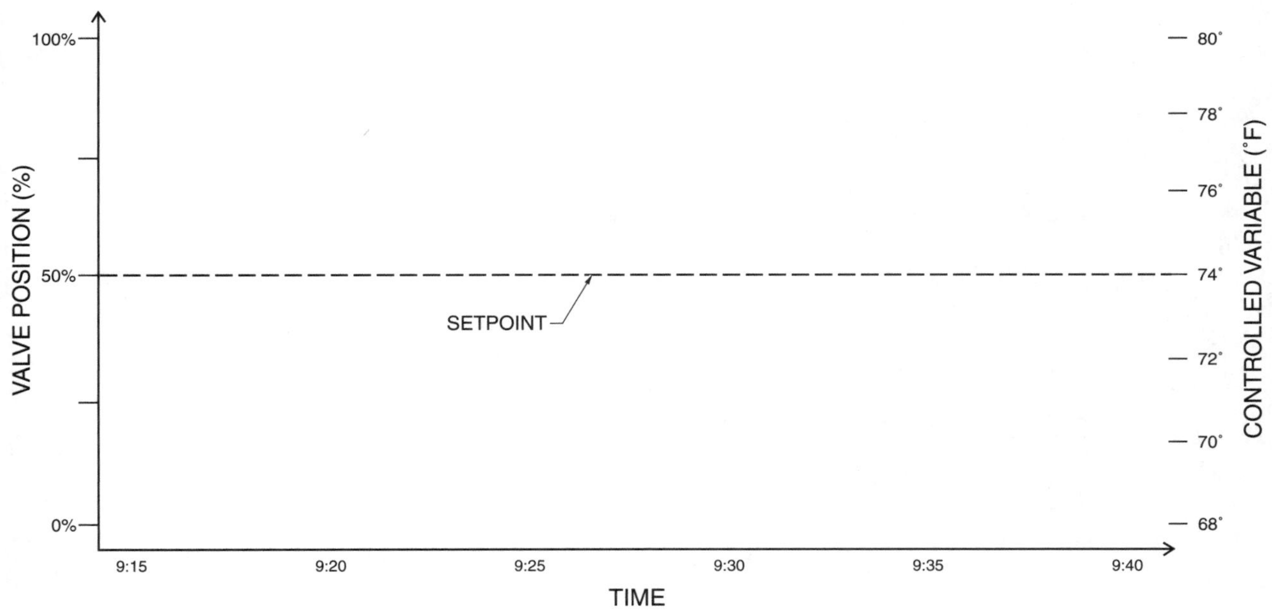

_______________ **8.** Is this level of control acceptable?

Section 7.1 Commercial Building Control Systems

REVIEW QUESTIONS

Name ______________________________ **Date** ____________________

Multiple Choice

__________ **1.** A(n) ___ control system is a control system in which compressed air is used to provide power for the control system.

A. self-contained
B. pneumatic
C. electrical
D. automated

__________ **2.** The control of the environment in commercial buildings is based on energy efficiency and ___.

A. cost
B. comfort
C. size of the building
D. type of building

__________ **3.** A pneumatic control system is a control system in which ___ is used to provide power for the control system.

A. fire
B. compressed air
C. electricity
D. natural gas

__________ **4.** In the 1960s, ___ came into common use and was adapted for use in commercial HVAC control systems.

A. fluid power
B. electricity
C. compressed air (pneumatic) power
D. solid-state, low-voltage direct current (DC) devices

__________ **5.** ___ is a designation of the contaminants present in the air.

A. Pollution
B. Indoor air quality
C. Uncirculated air
D. Makeup air

Completion

_______________ **1.** ___ is a designation of the contaminants present in the air of a building space.

_______________ **2.** ___ is a condition that occurs when people cannot sense a difference between themselves and the surrounding air.

_______________ **3.** Accurate control of the environment in a building space leads to ___ occupancy rates and profits.

_______________ **4.** Early control of the level of comfort in a living environment was done by ___.

REVIEW QUESTIONS

Name ______________________________ **Date** ______________

Multiple Choice

_______________ **1.** ___ control systems are used to control basic closed loop systems that require relatively low accuracy.
- A. Self-contained
- B. Pneumatic
- C. Electrical
- D. Automated

_______________ **2.** The power to a self-contained control system is supplied by a sealed, ___.
- A. duct
- B. fluid-filled element
- C. pump
- D. thermostatic expansion valve

_______________ **3.** As the temperature of the fluid in a pipe or air in a building space changes, the heat transfer in a self-contained control system changes the ___ of the fluid in the power element.
- A. pressure
- B. volume
- C. flow
- D. power output

_______________ **4.** The two most common applications of self-contained control systems are ___ and building space temperature control in steam or hot water heating systems.
- A. dampers
- B. air compressors
- C. thermostatic expansion valves
- D. room thermostats

_______________ **5.** Most self-contained control systems have relatively poor ___.
- A. power output
- B. accuracy
- C. pressure
- D. voltage signals

_______________ **6.** Pressure acts against a ___ that moves a valve body to regulate the flow of refrigerant, steam, hot water, or chilled water through the valve.
A. switch
B. fluid-filled element (bulb)
C. bimetallic element
D. diaphragm

Completion

_______________ **1.** The fluid-filled element of a self-contained control system is known as a(n) ___.

_______________ **2.** A self-contained control system is a control system that does not require a(n) ___.

_______________ **3.** The disadvantages of ___ control systems are that they cannot be expanded to provide sophisticated control sequences because only one setpoint adjustment is provided.

_______________ **4.** Self-contained control systems do not have diagnostic means available to troubleshoot or diagnose system failure, normally making ___ the only option.

Self-Contained Control System

_______________ **1.** adjustment stem

_______________ **2.** refrigerant outlet

_______________ **3.** diaphragm

_______________ **4.** fluid-filled element controls flow of refrigerant

_______________ **5.** valve open

_______________ **6.** liquid refrigerant inlet

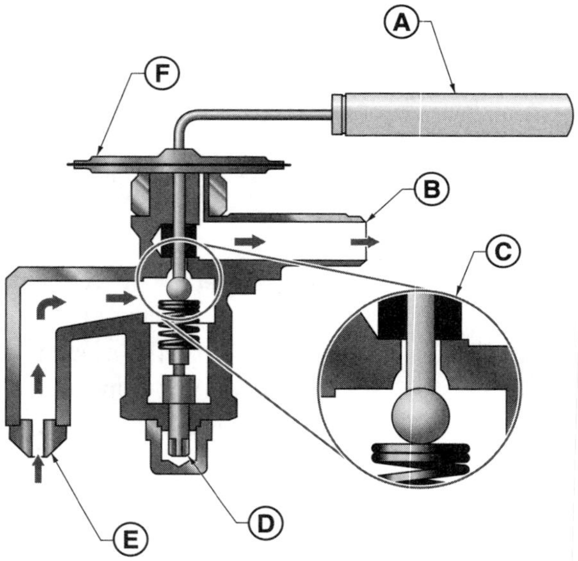

A
F
B
C
E
D

Section 7.3 Electrical Control Systems

REVIEW QUESTIONS

Name ______________________________ Date ____________________

Multiple Choice

______________ 1. Electrical control systems can use ___ to enable low-voltage control circuits to switch line-voltage devices.

A. relays
B. control devices
C. actuators
D. dampers

______________ 2. Humidity control is accomplished with a ___ that uses a moisture-sensing element to open or close a switch.

A. pressurestat
B. thermostat
C. humidistat
D. mercury bulb

______________ 3. ___ are often used in industrial and commercial systems in which access by occupants is not possible.

A. Relays
B. ON/OFF switches
C. Auxiliary devices
D. Line-voltage controls

______________ 4. Electrical control systems are not often used for ___ control because only digital (ON/OFF) control devices are used.

A. open loop
B. proportional
C. humidity
D. temperature

______________ 5. Humidity control is accomplished with a(n) ___ that uses a moisture-sensing element to open or close a switch.

A. humidistat
B. mercury bulb thermostat
C. electronic sensor
D. bulb sensor

Completion

____________________ **1.** A(n) ___ is a control system that uses electricity (24 VAC or higher) to operate the devices in the system.

____________________ **2.** A(n) ___ is a sensing device that consists of two different metals joined together.

____________________ **3.** ___ control devices are wired directly to the controlled devices such as compressors, fans, and pumps because both are at the same voltage level.

____________________ **4.** Modern thermostats use a(n) ___ to close or open a circuit to control the temperature in the space.

____________________ **5.** Thermostats are usually wired to a(n) ___ at the HVAC unit.

Section 7.4 Pneumatic Control Systems

REVIEW QUESTIONS

Name ______________________________ **Date** ______________________

Multiple Choice

_______________ **1.** Pneumatic control systems can be separated into ___ main groups of components based on their function.

A. two
B. three
C. four
D. five

_______________ **2.** ___ change or reroute the air supply from the transmitter or controller before it reaches the controlled devices.

A. Electric switches
B. Bleedports
C. Auxiliary devices
D. Dampers

_______________ **3.** A(n) ___ is an orifice that allows a small volume of air to be expelled to the atmosphere.

A. bleedport
B. exhaust port
C. damper
D. transmitter

_______________ **4.** Pneumatic control systems can be used in most control sequences because of their ___.

A. accuracy
B. power output
C. reliability
D. flexibility

_______________ **5.** A disadvantage of pneumatic control systems is that pneumatic air compressor stations require ___.

A. higher power supplies
B. regular maintenance
C. expensive piping and components
D. a single control sequence

Completion

______________ **1.** ___ devices are devices that are normally located between the transmitters or controllers and the controlled device.

______________ **2.** ___ control systems are rarely used in residential systems or packaged (rooftop or heat pump) units.

______________ **3.** Pneumatic control systems can be separated into ___ main groups of components based on their function.

Section 7.5 Electronic Control Systems

REVIEW QUESTIONS

Name ______________________ **Date** ______________

Multiple Choice

______ **1.** An electronic control system is a control system in which the power supply is ___ VDC or less.
A. 18
B. 24
C. 48
D. 115

______ **2.** A(n) ___ control system is a control system that uses a variable signal.
A. electrical
B. electronic
C. analog
D. digital

______ **3.** The result of an electronic control system is an output commonly between ___.
A. 10 VDC to 18 VDC
B. 24 VDC to 32 VDC
C. 110 VDC to 120 VDC
D. 220 VDC to 240 VDC

______ **4.** A resistive bridge circuit is a circuit containing ___.
A. two arms and two resistors
B. two arms and four resistors
C. four arms and two resistors
D. four arms and four resistors

______ **5.** The voltage signal is sent to a(n) ___, which opens or closes a valve or damper in response to the signal change.
A. switch
B. controller
C. relay
D. actuator

______ **6.** An advantage of electronic control systems is ___.
A. multiple control sequences
B. low cost
C. complexity
D. power output

Completion

_______________ 1. The resistive bridge circuit output value is ___ when the system is at setpoint.

_______________ 2. The disadvantage of ___ control systems is that they may require special diagnostic tools and procedures.

_______________ 3. ___ control systems are used primarily in large commercial buildings such as hospitals and schools.

_______________ 4. In an electronic control system, a signal from a sensor is wired to a(n) ___.

_______________ 5. The advantages of ___ control systems are that they are reliable, accurate, and relatively inexpensive.

Electronic Control System

_______________ 1. electronic heating or cooling valve

_______________ 2. 18 VDC out

_______________ 3. electronic room thermostat

_______________ 4. DC power supply

_______________ 5. 24 VAC in

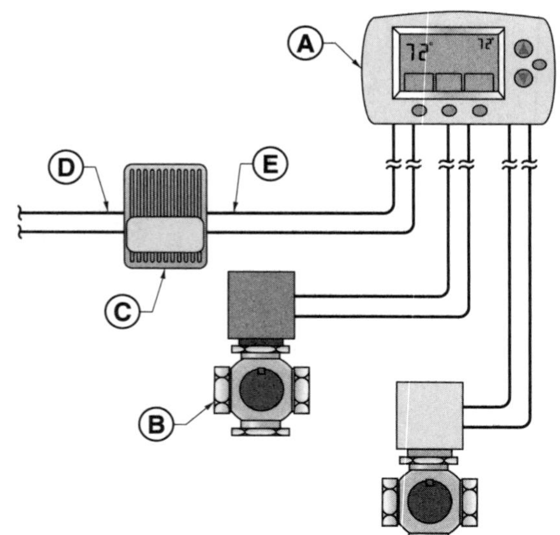

Section 7.6 Automated Control Systems

REVIEW QUESTIONS

Name ______________________________ **Date** ________________

Multiple Choice

________ **1.** A(n) ___ control system is a control system that uses digital solid-state components.
- A. electrical
- B. modulated
- C. pneumatic
- D. automated

________ **2.** ___ control systems are used in large commercial buildings such as hospitals, schools, military bases, and colleges.
- A. Automated
- B. Electrical
- C. Pneumatic
- D. Self-contained

________ **3.** In automated control systems, ___ are filtered to eliminate any interference or problems with the incoming power supply.
- A. transformers and rectifiers
- B. buses
- C. AC signals
- D. DC signals

________ **4.** In automated control systems, each controller has an individual ___ in its on-board memory.
- A. program
- B. power supply
- C. operator terminal
- D. VAV terminal box

________ **5.** The program parameters in automated control systems may be modified and new control sequences and options added by reprogramming the ___ with a desktop PC, notebook PC, portable operator terminal, or keypad display.
- A. actuators
- B. controller
- C. power signal
- D. power supply

__________ **6.** A disadvantage of automated control systems includes ___.
A. lack of versatility
B. lack of accuracy
C. poor security
D. high cost

Completion

__________ **1.** A(n) ___ control system uses ON/OFF (1/0) signals which represent numbers, setpoints, control sequences, etc.

__________ **2.** ___ control systems have become popular in commercial buildings because of their increased reliability and the increase in the power and capacity of personal computers (PCs).

__________ **3.** A building automation system (BAS) is a system that uses ___ to control the energy-using devices in a building.

Section 7.7 System-Powered Control Systems

REVIEW QUESTIONS

Name ______________________________ **Date** ______________________

Multiple Choice

______________ **1.** A system-powered control system is a control system in which the ___ is used as the power supply.
- A. steam or hot water
- B. electricity
- C. fluid-filled element
- D. duct pressure

______________ **2.** In a system-powered control system, the air pressure, measured at about ___″ wc, powers the control system.
- A. 1
- B. 10
- C. 100
- D. 1000

______________ **3.** In system-powered control systems, the bimetallic element moves in response to changes in the ___.
- A. duct pressure
- B. building space temperature
- C. bellows
- D. control sequences

______________ **4.** System-powered control systems allow flexibility of zoning in a building because ___.
- A. pressure is easily diverted with the use of dampers
- B. no piping or wiring needs to be changed
- C. zoning can be changed with remote access controllers
- D. VAV terminal boxes can divert air in different directions

______________ **5.** System-powered control systems have reduced installation time because controls are normally ___.
- A. pre-calibrated
- B. remotely calibrated
- C. factory mounted
- D. field mounted

Completion

______________________ **1.** A(n) ___ control system is a control system in which the duct pressure developed by the fan system is used as the power supply.

______________________ **2.** System-powered control systems cannot be adapted to other types of control such as ___.

______________________ **3.** ___ control systems were developed to avoid the installation costs of pneumatic piping runs from an air compressor station.

Section 7.8 Hybrid Control Systems

REVIEW QUESTIONS

Name ______________________________ **Date** ________________

Multiple Choice

______________ **1.** In ___ control systems, transducers are used as an interface between different control system technologies.
A. electrical
B. electronic
C. hybrid
D. pneumatic

______________ **2.** Hybrid control systems are common in ___ building control.
A. residential
B. commercial
C. single-story industrial
D. multiple-story industrial

______________ **3.** Hybrid control systems contain multiple control technologies requiring different ___.
A. offsets
B. system components
C. system pressures
D. power supplies

______________ **4.** In a hybrid control system, the transducer has an input signal of ___.
A. 0 VDC to 10 VDC
B. 10 VDC to 20 psig
C. 0 psig to 20 psig
D. 10 psig to 20 VDC

______________ **5.** Hybrid control systems minimize cost by ___.
A. decreasing the number of components used in the system
B. maximizing power output
C. reusing old components and installed materials
D. using less expensive power sources

Completion

_______________ **1.** The most common application of a(n) ___ system is the retrofit of an automated control system to a pneumatic control system.

_______________ **2.** In a hybrid system, ___ difficulty may be increased due to system complexity.

_______________ **3.** A(n) ___ control system is a control system that uses multiple control technologies.

_______________ **4.** ___ control systems minimize cost by reusing old components and installed materials such as control cabinets, wiring, and piping.

Name ______________________________ Date ______________

Activity 7-1. Control System Identification

A new firm has taken over the service contract on a building. Before service can be performed, knowledge of the control system must be obtained. Use the air handling unit drawing to answer the questions.

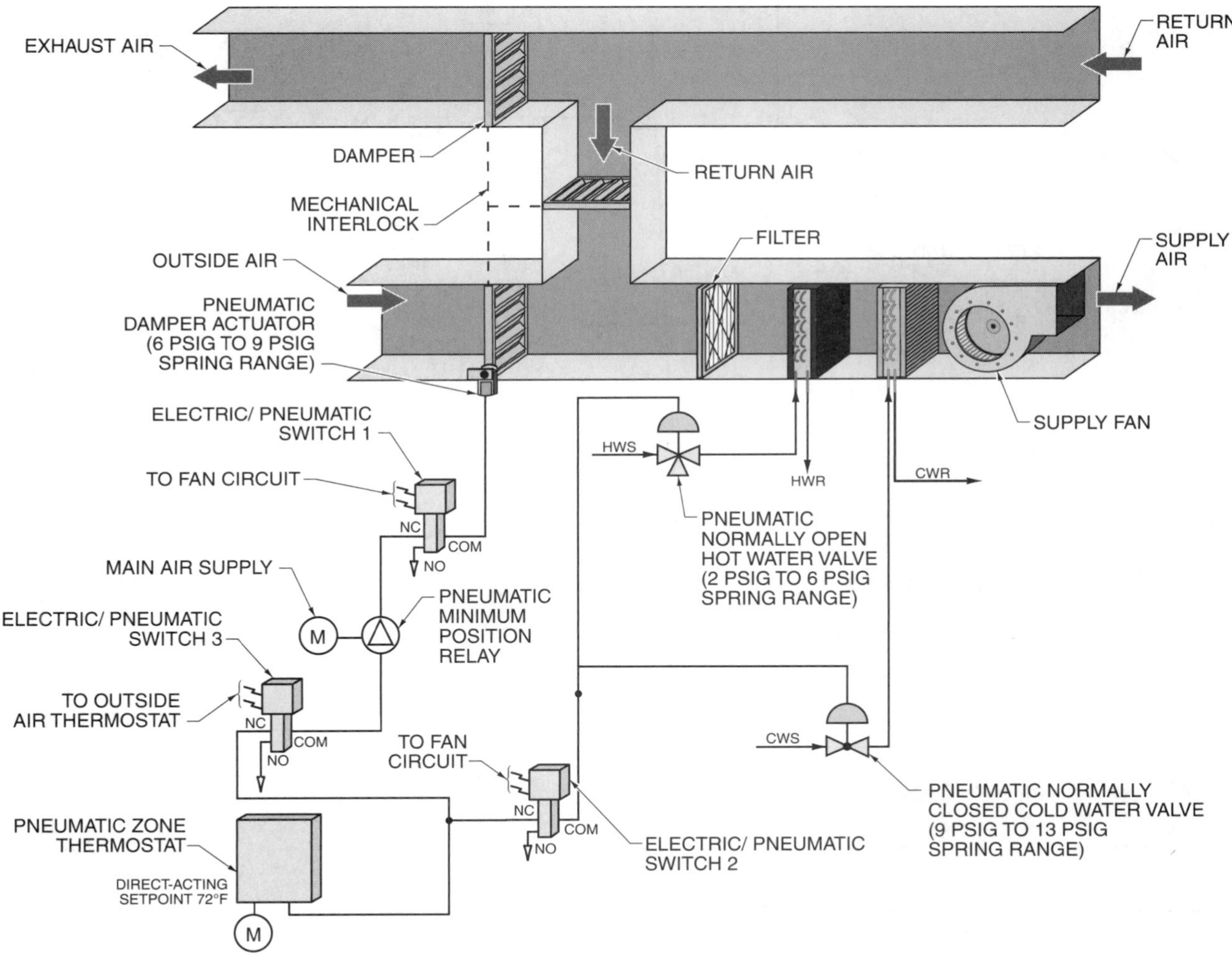

__________ **1.** What type of air handling unit is it?

__________ **2.** Based on the drawing, what type of control system is it?

Activity 7-2. Control System Device Identification

Use the pneumatic control system drawing as a guide to answer the questions on the air handling unit drawing from Activity 7-1.

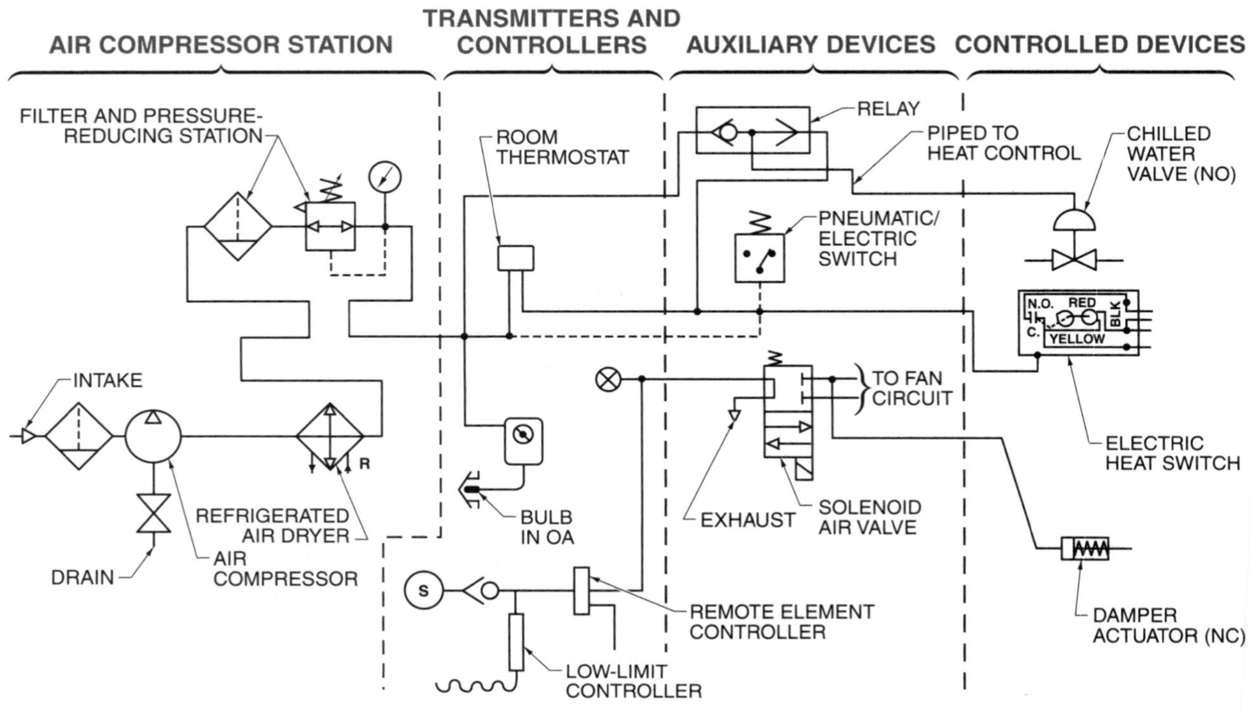

1. List the transmitter(s)/controller(s) in the air handling unit.

____

____

2. List the auxiliary device(s) in the air handling unit.

____

____

3. List the controlled device(s) in the air handling unit.

____

____

____________________ **4.** The setpoint of the room thermostat is ___°F.

____________________ **5.** What device on the unit reduces indoor air quality problems?

____________________ **6.** Is a humidity control device included on the unit?

7. What items require replacement on a regular basis?

____

____

____________________ **8.** What type of water is used in the unit?

____________________ **9.** What devices supply water to the unit?

____________________ **10.** What device moves air in the system?

Section 8.1 Air Compressor Stations

REVIEW QUESTIONS

Name ______________________ Date ______________

Multiple Choice

______ **1.** An ___ filter is a device that removes oil droplets from a pneumatic system by forcing compressed air to change direction quickly.

A. oil removal
B. air line
C. intake air
D. oil saturated

______ **2.** A(n) ___ drier is a device that uses refrigeration to lower the temperature of compressed air.

A. refrigerated air
B. desiccant
C. charcoal
D. air line

______ **3.** A ___ compressor adds kinetic energy to accelerate air and convert the velocity energy to pressure energy with a diffuser.

A. reciprocating
B. centrifugal
C. rotary
D. dynamic

______ **4.** As the compressor run time increases, ___, affecting the motor and the compressor.

A. resistance increases
B. heat builds
C. parts wear
D. volume compresses

______ **5.** Pressure drop in a pipe is reduced by ___.

A. decreasing pipe diameter as the length increases
B. decreasing pipe diameter as the length decreases
C. increasing pipe diameter as the length increases
D. increasing pipe diameter as the length decreases

______ **6.** A(n) ___ is a device that consists of a housing containing a centrifugal deflector plate and a small filtration element.

A. automatic drain
B. air line filter
C. desiccant air dryer
D. oil remover filter

______________ **7.** A ___ compressor is a compressor that compresses a fixed amount of air with each cycle.
A. rotary
B. vane
C. centrifugal
D. positive-dsplacement

______________ **8.** On the suction stroke of a reciprocating compressor, the suction valve (inlet valve) opens, and ___ is drawn into the cylinder.
A. heat
B. oil carryover
C. air from the atmosphere
D. refrigerant

______________ **9.** A(n) ___ is commonly located in a metal housing at the intake of the air compressor.
A. desiccant drier
B. intake air filter
C. manual drain
D. centrifugal compressor

______________ **10.** A ___ compressor is a compressor that uses centrifugal force to compress air.
A. centrifugal
B. dynamic
C. reciprocating
D. scroll

Completion

______________ **1.** A(n) ___ compressor is a positive-displacement compressor that has multiple vanes located in an offset rotor.

______________ **2.** A(n) ___ is a unit of measure equal to 0.000039″.

______________ **3.** ___ is the adhesion of a gas or liquid to the surface of a porous material.

______________ **4.** A(n) ___ is a valve that restricts and/or blocks downstream airflow.

______________ **5.** A(n) ___ valve is a device that prevents excessive pressure from building up by venting air to the atmosphere.

______________ **6.** A(n) ___ is a device that is opened and closed manually to drain moisture from the receiver.

______________ **7.** A(n) ___ compressor compresses a fixed quantity of air with each cycle.

______________ **8.** The condition of a filtration element is determined by measuring the ___ across the element.

______________ **9.** A(n) ___ is a device that removes moisture by adsorption.

______________ **10.** On the ___ stroke, the suction valve is pushed closed, compressing the air and discharging it at a high pressure through the discharge valve to the receiver.

__________ **11.** A(n) ___ is a device that is normally piped from the lowest part of the receiver and opens based on differential pressure or moisture level build-up.

__________ **12.** ___ is lubricating oil that leaks by the piston rings and is carried into the compressed air system.

__________ **13.** Air ___ convert the mechanical energy provided by an electric motor into the potential energy of compressed air.

__________ **14.** A(n) ___ is a device that opens and closes automatically at a predetermined interval to drain moisture from the receiver.

__________ **15.** A ___ compressor is a compressor that contains a pair of screw-like helical gears that interlock as they rotate.

__________ **16.** An air compressor is a component that takes air from the ___ and compresses it to increase its pressure.

Oil Removal Filters

__________ **1.** automatic drain

__________ **2.** bowl guard

__________ **3.** filtration element

__________ **4.** body

__________ **5.** bowl O-ring

__________ **6.** nut

__________ **7.** transparent bowl

__________ **8.** differential pressure indicator

__________ **9.** gasket

__________ **10.** O-ring

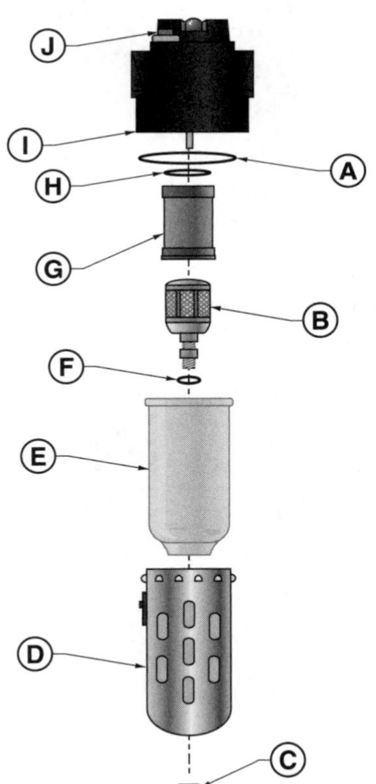

Air Compressor Schematic Symbol Identification

______________ **1.** filter/separator with manual drain

______________ **2.** automatic drain

______________ **3.** pressure switch

______________ **4.** intake air filter

______________ **5.** safety relief valve

______________ **6.** pressure regulator with gauge

______________ **7.** filter/separator with automatic drain

______________ **8.** air compressor

______________ **9.** air drier

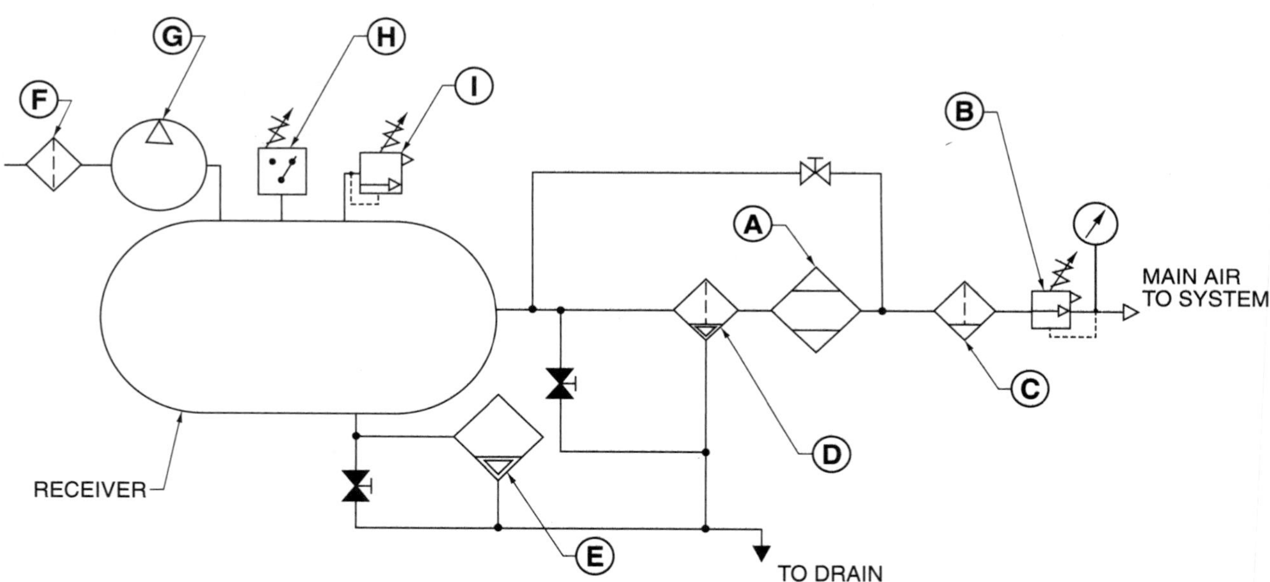

Section 8.2 Installation and Maintenance

REVIEW QUESTIONS

Name ______________________________ **Date** ______________________

Multiple Choice

______________ **1.** In general, air compressors should use ___ oil.

A. synthetic
B. detergent
C. non-detergent
D. water-based

______________ **2.** ___ tubing is often used in electrical conduits to protect the tubing from harsh conditions.

A. Poly
B. Surgical
C. Copper
D. Rubber

______________ **3.** A(n) ___ test determines whether an air compressor is sized properly for the job.

A. pressure
B. starts per hour
C. duty cycle
D. run time

______________ **4.** A pressure test should be performed and recorded ___.

A. daily
B. monthly
C. quarterly
D. annually

______________ **5.** A(n) ___ is a device that operates one compressor during one pumping cycle and the other compressor during the next pumping cycle.

A. lead/lag switch
B. compressor alternator
C. alternator
D. equalizing compressor

Completion

______________ **1.** A(n) ___ test is an air compressor performance test that measures the percentage of time that a compressor runs to maintain a supply of compressed air to the control system.

______________ **2.** A(n) ___ compressor consists of two air compressors and two electric motors on one common receiver.

_______________ 3. In an air loop configuration, the compressed air enters both ends of the loop, reducing potential ___ over long runs.

_______________ 4. The ___ is found by dividing the compressor on time by the compressor on time plus the compressor off time.

_______________ 5. A(n) ___ is a pressure switch that determines which compressor is the primary (lead) compressor and which compressor is the backup (lag) compressor.

_______________ 6. If compressed air that contains ___ is exposed to low outside temperatures, air lines may freeze and split.

Air Compressor Preventive Maintenance Procedure

Identify the procedures shown.

_______________ 1. Properly replace belt guard and turn power ON to compressor.

_______________ 2. Drain tank. Check volume of oil in water.

_______________ 3. Turn power OFF to compressor. Follow lockout/tagout procedures.

_______________ 4. Check oil level in crankcase.

_______________ 5. Remove belt guard. Check belt for cracking, glazing, and tension.

_______________ 6. Manually operate safety relief valves.

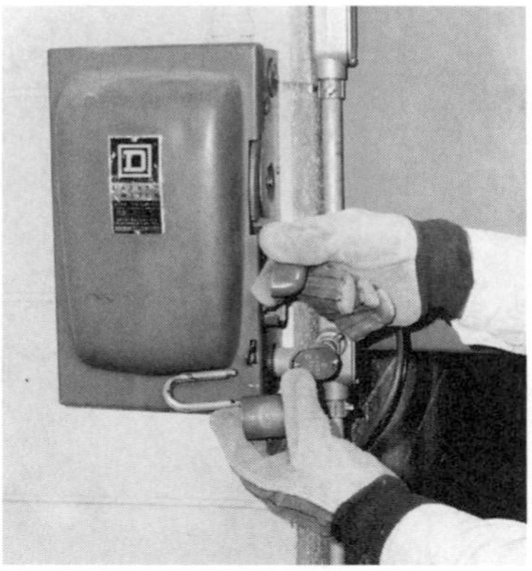

Ⓐ Ⓑ Ⓒ

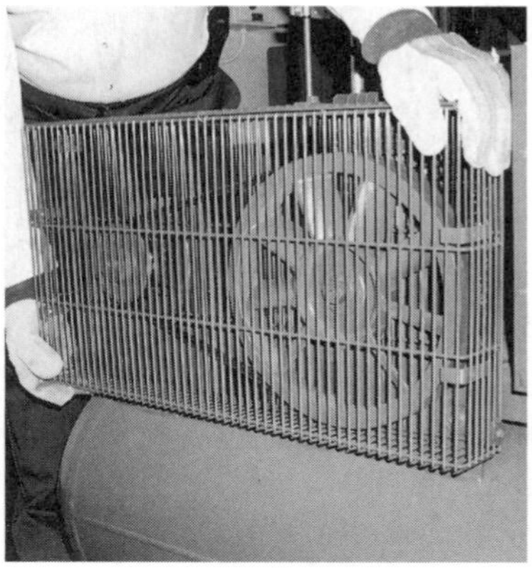

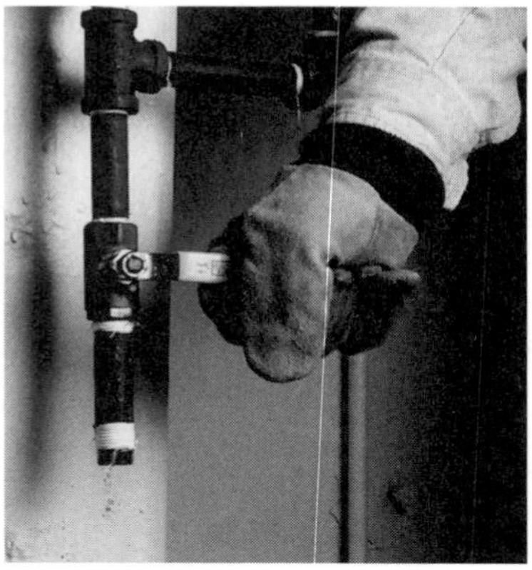

Ⓓ Ⓔ Ⓕ

Name ______________________________ **Date** ____________________

Activity 8-1. Air Compressor Performance

A control air compressor under normal load is timed as running from 7:10 AM to 7:15 AM. The compressor then starts again at 7:25 AM.

__________ **1.** The compressor run time is ___%.

__________ **2.** Is the compressor run time within normal parameters?

__________ **3.** The compressor has ___ starts per hour.

__________ **4.** Is the compressor number of starts per hour within normal parameters?

Activity 8-2. Log Sheet Completion

Complete the air compressor preventive maintenance log.

1.

AIR COMPRESSOR PREVENTIVE MAINTENANCE LOG

Air Compressor Location:	Number:	Date:	On-Time:	Off-Time:	Run-Time:	Starts/Hour:	PM Performed?	Problems:	Notes:	Initials:
Admin Bldg	1	11/7	5 min	10 min			Yes	None		
Admin Bldg	2	11/7	4 min	4 min			Yes	None		
Gym	1	11/7	4 min	10 min			Yes	None		

Activity 8-3. Control Air System Component Connection

Connect the devices in the control air system.

1.

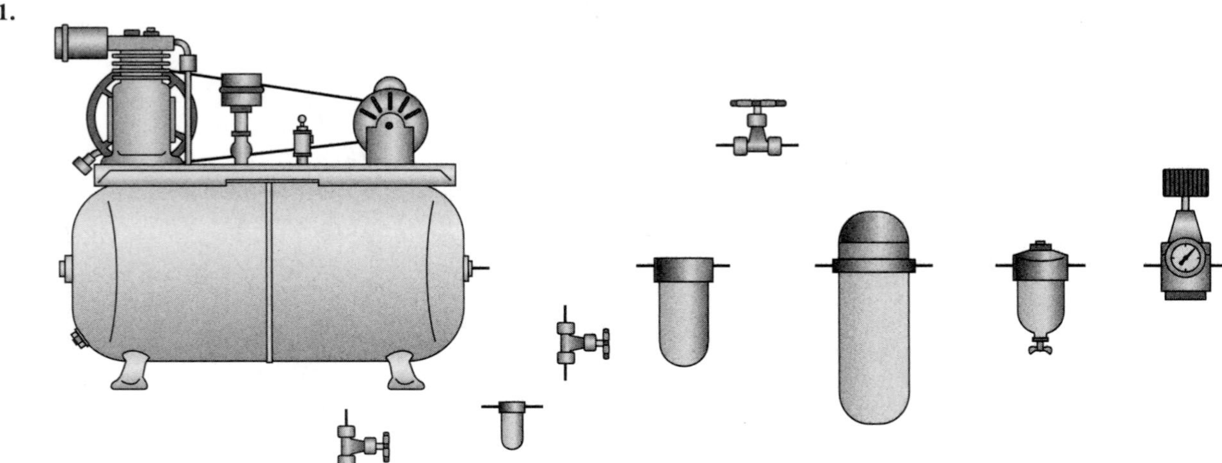

Section 9.1 Actuators

REVIEW QUESTIONS

Name ______________________________ **Date** ____________________

Multiple Choice

__________ **1.** A(n) ___ is a flexible device that transmits the force of the incoming air pressure to the piston cup and then to the spring and shaft assembly.
A. diaphragm
B. piston cup
C. aluminum end cap
D. spring and shaft assembly

__________ **2.** Actuators may be ___ or valve actuators.
A. damper
B. control
C. duplex
D. pump

__________ **3.** On a pneumatic actuator, the end cap with an air fitting is normally made of ___.
A. inexpensive cardboard
B. rubber or another flexible plastic
C. aluminum or high-impact plastic
D. corrosion-resistant steel

__________ **4.** A ___ is a device that transfers the force generated by the air pressure against the diaphragm to the spring and shaft assembly.
A. valve actuator
B. damper
C. spring range
D. piston cup

__________ **5.** Diaphragm force is found by ___.
A. multiplying air pressure from the controller by area of the diaphragm
B. dividing air pressure from the controller by area of the diaphragm
C. multiplying air pressure from the controller by area of the actuator
D. dividing air pressure from the controller by area of the actuator

__________ **6.** The ___ converts the air pressure change at the diaphragm into mechanical movement.
A. piston
B. valve actuator
C. spring and shaft assembly
D. damper

_______________ **7.** ___ is the difference in pressure at which an actuator shaft moves and stops.
A. Force
B. Spring range
C. Kinetic pressure
D. Spring range shift

_______________ **8.** The ___ is attached to the actuator and compressed, adding air to the actuator until its stroke is complete.
A. squeeze bulb
B. spring
C. damper
D. diaphragm

Completion

_______________ **1.** A(n) ___ is a device that accepts a signal from a controller and causes a proportional mechanical motion to occur.

_______________ **2.** A(n) ___ is a device that transfers the force generated by the air pressure against the diaphragm to the spring and shaft assembly.

_______________ **3.** Spring range ___ is a condition in which actuators with different spring ranges interfere with each other.

_______________ **4.** ___ is the difference in pressure at which an actuator shaft moves and stops.

_______________ **5.** ___ occurs in systems that have heating and cooling systems that have closely coordinated spring ranges in order to obtain close sequencing.

Section 9.2 Dampers

REVIEW QUESTIONS

Name ______________________________ **Date** ____________________

Multiple Choice

__________ **1.** ___ blade dampers are the most common damper used in HVAC systems.
A. Round
B. Opposed
C. Parallel
D. Elliptical

__________ **2.** Dampers are classified as parallel, ___, and round blade dampers.
A. dual air-stream
B. square
C. opposed
D. linear

__________ **3.** ___ blade dampers are less expensive and provide better air mixture when installed to cause outside air and return air to collide.
A. Opposed
B. Round
C. Elliptical
D. Parallel

__________ **4.** Opposed blade dampers are used in applications that require two air streams to mix in order to prevent ___.
A. chemical imbalances
B. freeze up
C. hot spots
D. uneven air flow

__________ **5.** ___ blade dampers have a nonlinear flow characteristic but are widely used in small terminal units such as VAV terminal boxes.
A. Elliptical
B. Opposed
C. Round
D. Parallel

__________ **6.** A(n) ___ type lubricant is often used to lubricate dampers at their lubrication points.
A. dry, graphite
B. thick, viscous
C. oil or oil-like
D. dry, chalky

Completion

______________ **1.** A(n) ___ is an adjustable metal blade or set of blades used to control the flow of air.

______________ **2.** The frequency of damper ___ depends on the unit run hours and damper service.

______________ **3.** Dampers are normally constructed from ___ that is treated to resist corrosion and rust.

Damper Seals

______________ **1.** bottom seal

______________ **2.** end seal

______________ **3.** damper blade

______________ **4.** damper blade seal

______________ **5.** damper blade pin

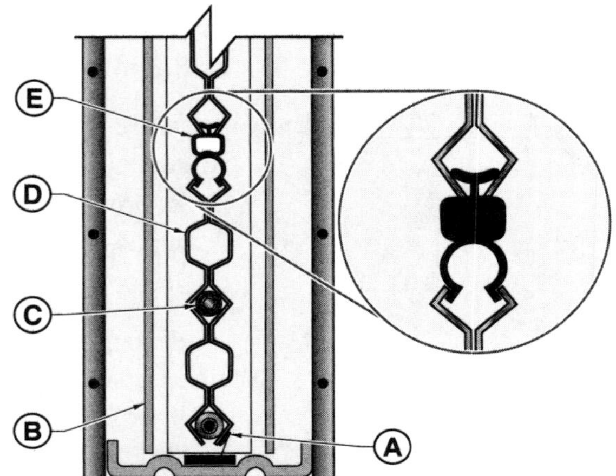

Damper Configurations

______________ **1.** opposed blade

______________ **2.** round blade

______________ **3.** parallel blade

Jackson Systems, LLC

Ⓐ

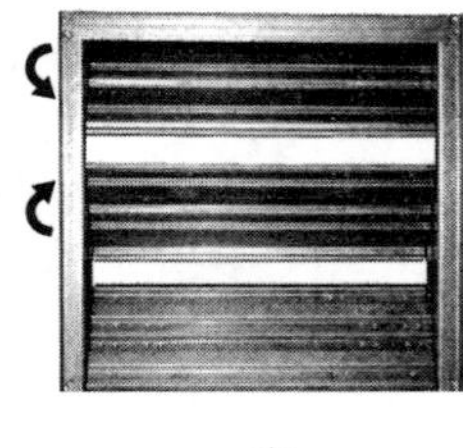

Ⓑ

Jackson Systems, LLC

Ⓒ

Section 9.3 Valves

REVIEW QUESTIONS

Name ______________________________ **Date** ____________________

Multiple Choice

__________ **1.** A(n) ___ valve is a three-way valve that has two inlet ports and one outlet port.
A. equal percentage
B. diverting
C. butterfly
D. mixing

__________ **2.** A(n) ___ valve is a valve in which the flow through the valve is equal to the amount of valve stroke.
A. butterfly
B. linear
C. quick-opening
D. diverting

__________ **3.** Valve ___ is the relationship between the maximum flow and the minimum controllable flow through a valve.
A. turn down ratio
B. shut-off rating
C. control ratio
D. shut-down rating

__________ **4.** ___ is used in cooling towers or water source heat pumps.
A. Cool water
B. Warm water
C. Hot water
D. Steam

__________ **5.** A valve ___ is a valve component that transmits the force of the actuator to the valve plug.
A. stem
B. body
C. disc
D. packing

__________ **6.** ___ valves have a plug on the bottom of the valve body that enables the disassembly and removal of the disc and other internal parts.
A. Linear
B. Mixing
C. Normally open
D. Normally closed

______________ **7.** Pump or system ____ are installed to maintain a relatively constant supply and return pressure, which reduces pump wear.
A. mixing valves
B. relief valves
C. valve discs
D. outlets

______________ **8.** A(n) ___ valve is a valve with a round plate that rotates to control flow.
A. equal percentage
B. quick-opening
C. butterfly
D. two-way

______________ **9.** ___ is valve erosion caused by steam molecules being accelerated by a valve operating in a cracked-open position for a long period of time.
A. Wire drawing
B. Seal cracking
C. Wear
D. Rusting

______________ **10.** ___ is the gallons per minute of flow through a valve at a 1 psig pressure drop across the valve.
A. P_v
B. G_v
C. C_v
D. D_v

______________ **11.** Valves that have unstable performance at minimum flow may indicate that the ___ is being exceeded.
A. valve turn down ratio
B. safety rating
C. manufacturer rating
D. spring range

______________ **12.** Valve ___ is performed when the valve begins to leak excessively at the valve stem and packing gland assembly.
A. testing
B. repacking
C. erosion
D. certification

______________ **13.** Valve seat erosion can be corrected by ___ the valve seat.
A. repacking
B. rebuilding
C. resealing
D. grinding

Completion

_______________ **1.** A(n) ___ is a device that controls the flow of fluids in an HVAC system.

_______________ **2.** A(n) ___ valve has two pipe connections.

_______________ **3.** A(n) ___ valve does not allow fluid to flow when the valve is in its normal position.

_______________ **4.** A(n) ___ is a valve component that consists of a metal shaft, normally made of stainless steel, that transmits the force of the actuator to the valve plug.

_______________ **5.** ___ characteristics is the relationship between the valve stroke and flow through the valve.

_______________ **6.** ___ is a bulk deformable material or one or more mating deformable elements reshaped by manually adjustable compression.

_______________ **7.** Valve ___ is performed when a valve is mechanically damaged or the internal parts are worn by foreign material or wire drawing.

_______________ **8.** A(n) ___ valve is a valve that allows fluid to flow when the valve is in its normal position.

_______________ **9.** Steam valves are always ___ valves because steam does not require a return to the steam header or boiler.

Valve Rebuild Kit

_______________ **1.** stem and disc holder assembly

_______________ **2.** follower spring

_______________ **3.** packing nut

_______________ **4.** packing lubricant

_______________ **5.** follower

_______________ **6.** valve body

_______________ **7.** plug

_______________ **8.** disc spring

_______________ **9.** packing gland

_______________ **10.** packing

_______________ **11.** bonnet

_______________ **12.** disc

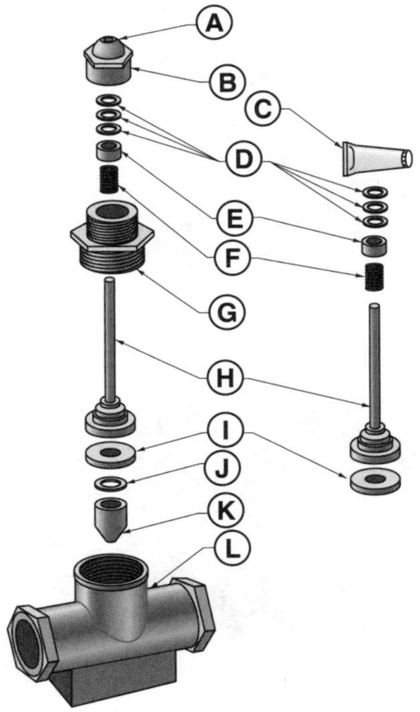

A
B
C
D
E
F
G
H
I
J
K
L

Valve Packing

_______________ 1. packing gland

_______________ 2. packing nut

_______________ 3. bonnet

_______________ 4. packing

_______________ 5. valve stem

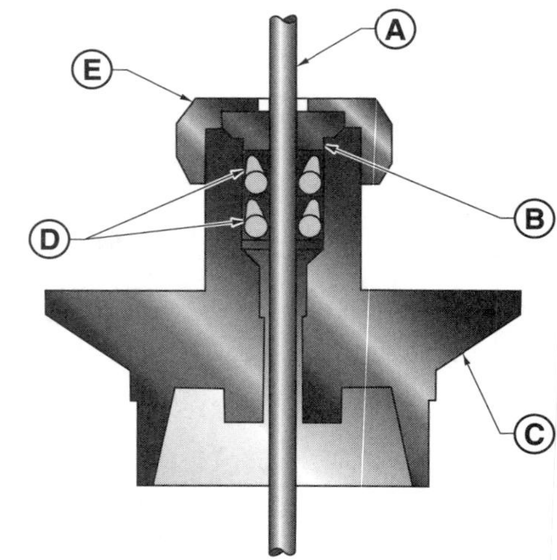

HVAC Control Valve Components

_______________ 1. valve body

_______________ 2. packing gland

_______________ 3. air fitting

_______________ 4. piston cup

_______________ 5. stem

_______________ 6. valve plug

_______________ 7. seat

_______________ 8. disc

_______________ 9. flange

_______________ 10. packing nut

_______________ 11. valve port

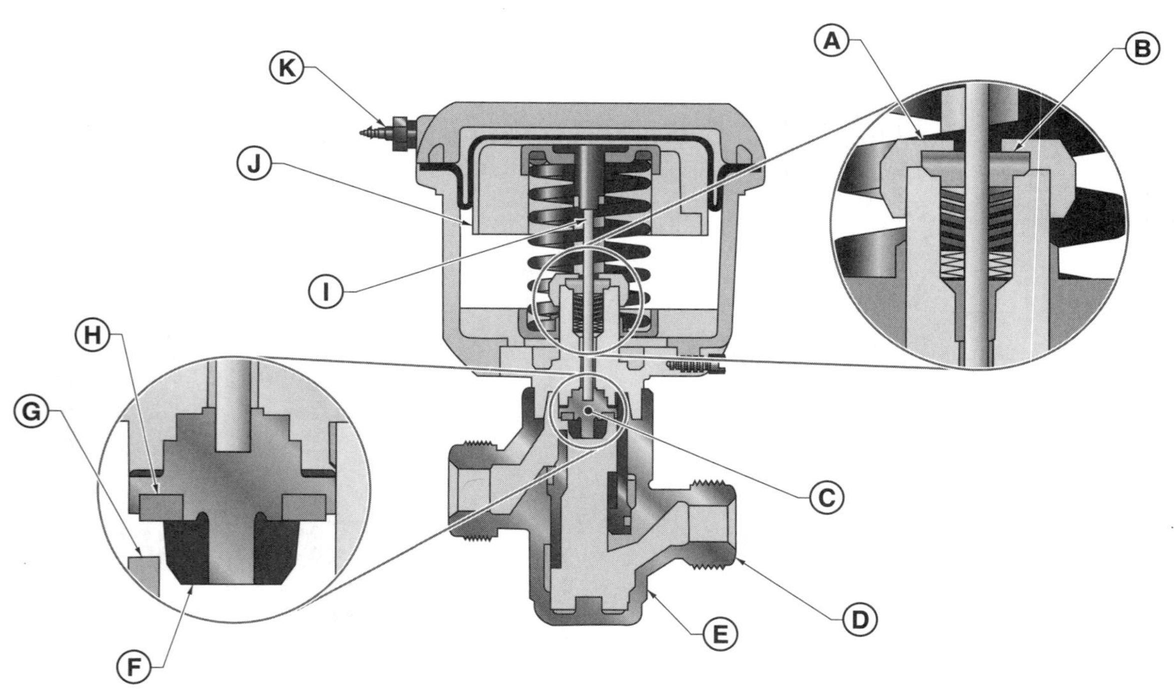

Section 9.4 Normally Open and Normally Closed Dampers and Valves

REVIEW QUESTIONS

Name ______________________________ **Date** ______________________

Multiple Choice

_______________ **1.** The position of a damper or valve with no air pressure on the actuator is referred to as the ___ position of the actuator.
- A. starting
- B. fail-safe
- C. normally open
- D. normally closed

_______________ **2.** A computer room cooling system has dampers and valves that ___, regardless of the climate, because computer rooms require extensive cooling.
- A. are normally closed
- B. are normally open
- C. fail closed
- D. fail open

_______________ **3.** Most air handling systems use normally closed outside air dampers because ___.
- A. indoor air requires specific ratios of recirculated air and outside air
- B. outside air contains pollutants
- C. indoor air should be separated from outside air with filters
- D. outside summer and winter temperatures can be extreme

_______________ **4.** Normally open and normally closed damper operation is determined by the attachment of the ___ to the damper blade.
- A. hot water return
- B. pump
- C. linkage
- D. air duct

_______________ **5.** A three-way valve is normally open if it is piped to have water flow through the coil if the ___.
- A. air pressure to the actuator is increased
- B. outside air damper is closed
- C. outside air damper is open
- D. air pressure to the actuator is removed

Completion

_______________ **1.** Normally closed dampers and valves fail to the ___ position if the air pressure is removed from the damper or valve actuator.

_______________ **2.** Cooling system dampers and valves in southern climates normally fail ___ to flow to allow continued cooling if a problem occurs.

_______________ **3.** ___ are used to prevent continual discharge of steam or water into the air stream and prevent dangerous levels of humidity in a building.

_______________ **4.** A(n) ___ damper or valve allows fluid to flow when the damper or valve is in its normal position.

_______________ **5.** ___ outside air dampers are normally used because the outside air causes excessively cold conditions in the winter.

Pneumatic Actuators, Dampers, and Valves

ACTIVITIES

Name ______________________ **Date** ______________

Activity 9-1. Valve Data Sheet

Answer the questions using the valve data sheet.

VALVE DATA SHEET

Product			VP8200 Series Bronze Control Valves
Service*			Hot Water, Chilled Water, Glycol Solutions, or Steam for HVAC Systems
Valve Body Size/Cv	**1/2 in.**		0.73, 1.8, and 4.6
	3/4 in.		7.3
	1 in.		11.6
	1-1/4 in.		18.5
	1-1/2 in.		28.9
	2 in.		46.2
Valve Stroke			5/16 in. for 1/2 and 3/4 in. Valves
			1/2 in. for 1 and 1-1/4 in. Valves
			3/4 in. for 1-1/2 and 2 in. Valves
Valve Body Rating Valve Assembly Maximum Allowable Pressure/Temperature	**Steam**	**Brass Trim**	38 psig Saturated Steam at 284°F
		SS Trim	100 psig Saturated Steam at 284°F
	Water	**Brass Trim**	400 psig Up to 150°F, Decreasing to 365 psig at 248°F
		SS Trim	400 psig Up to 150°F, Decreasing to 308 psig at 338°F
Inherent Flow Characteristics			Equal Percentage: N.O./PDTC and N.C./PDTO Valves
			Linear: Three-Way Mixing Valves
Rangeability**			25:1 for All Sizes
Spring Range Pneumatic Actuators			3 to 6 psig: 3 to 7 psig for MP6000 Series
			4 to 8 psig
			9 to 13 psig
Maximum Recommended Operating Pressure Drop	**Steam**	**Brass Trim**	15 psig for All Valve Sizes
		SS Trim	100 psig for All Valve Sizes
	Water	**All Trim**	35 psig for 1/2 through 1-1/2 in. Valves
			30 psig for 1-1/2 and 2 in. Valves
Maximum Actuator Supply Pressure			25 psig Maximum

*Proper water treatment is recommended.
** Rangeability is the ratio of maximum flow to minimum controllable flow.

______________ **1.** The ___ valve should be selected if an application requires a Cv of 25.

______________ **2.** The valve stroke is ___″.

______________ **3.** The three-way mixing valves have ___ flow characteristics.

______________ **4.** The maximum recommended operating pressure drop for a 1½″ water valve is ___ psig.

5. List all possible spring ranges for pneumatic actuators.

____

____

____

Activity 9-2. Hot Water Heat Exchanger

Answer the questions using the heat exchanger drawing.

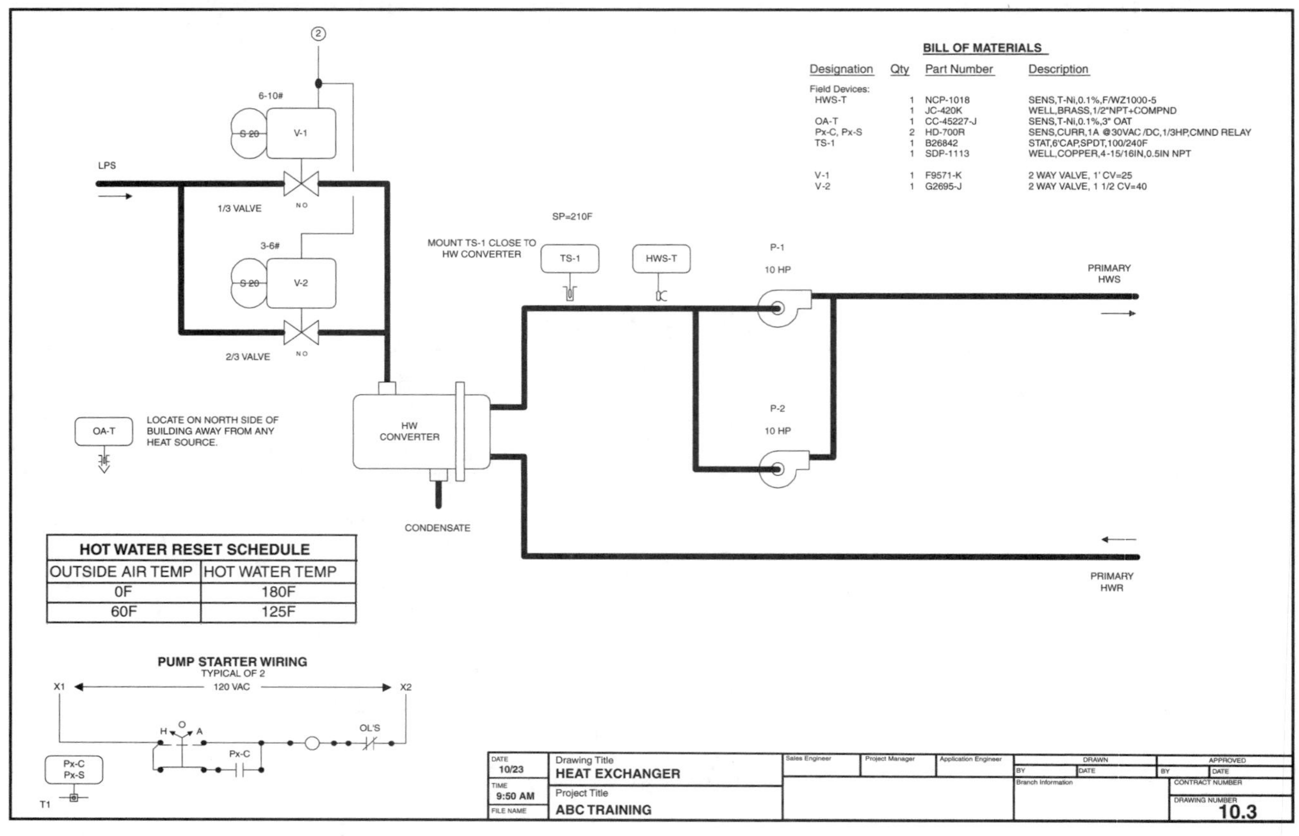

________________ **1.** The type of application is ___.

2. Are valves V-1 and V-2 normally open or normally closed? Why?

____

____

________________ **3.** Are valves V-1 and V-2 the same size?

________________ **4.** Valve V-1 is a(n) ___ valve.

________________ **5.** The Cv of valve V-1 is ___.

6. Why do the actuators for valve V-1 and valve V-2 have different spring ranges?

____

____

________________ **7.** According to the hot water reset schedule, if the outside air temperature is 60°F, the hot water setpoint is ___°F.

8. Would valve V-2 tend to be open or closed at a 55°F outside air temperature? Why?

____

____

9. If the pumps were 20 HP instead of 10 HP and delivered twice the amount of water and flow rate, would it affect the valve sizing? Why or why not?

____

____

Activity 9-3. Air Handling Unit

Answer the questions using the air handling unit drawing.

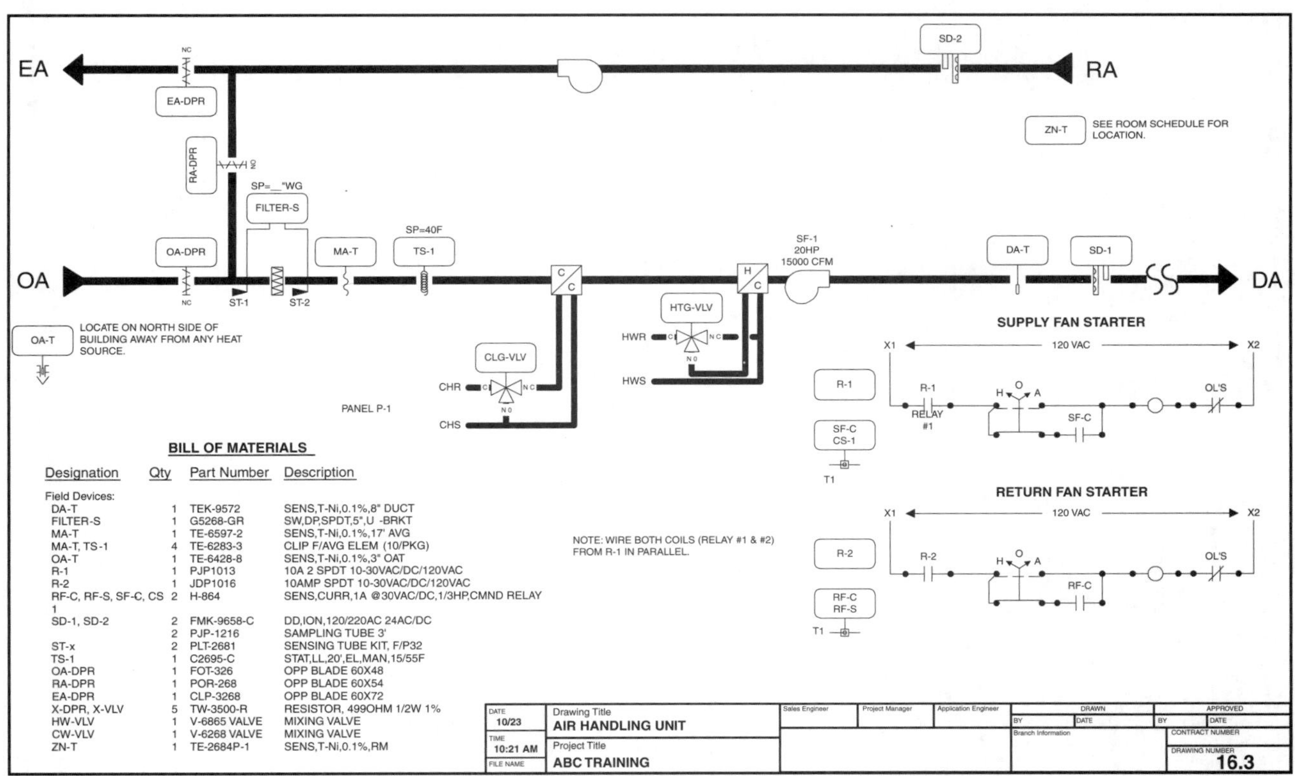

BILL OF MATERIALS

Designation	Qty	Part Number	Description
Field Devices:			
DA-T	1	TEK-9572	SENS,T-Ni,0.1%,8" DUCT
FILTER-S	1	G5268-GR	SW,DP,SPDT,5",U -BRKT
MA-T	1	TE-6597-2	SENS,T-Ni,0.1%,17' AVG
MA-T, TS-1	4	TE-6283-3	CLIP F/AVG ELEM (10/PKG)
OA-T	1	TE-6428-8	SENS,T-Ni,0.1%,3" OAT
R-1	1	PJP1013	10A 2 SPDT 10-30VAC/DC/120VAC
R-2	1	JDP1016	10AMP SPDT 10-30VAC/DC/120VAC
RF-C, RF-S, SF-C, CS 1	2	H-864	SENS,CURR,1A @30VAC/DC,1/3HP,CMND RELAY
SD-1, SD-2	2	FMK-9658-C	DD,ION,120/220AC 24AC/DC
	2	PJP-1216	SAMPLING TUBE 3'
ST-x	2	PLT-2681	SENSING TUBE KIT, F/P32
TS-1	1	C2695-C	STAT,LL,20',EL,MAN,15/55F
OA-DPR	1	FOT-326	OPP BLADE 60X48
RA-DPR	1	POR-268	OPP BLADE 60X54
EA-DPR	1	CLP-3268	OPP BLADE 60X72
X-DPR, X-VLV	5	TW-3500-R	RESISTOR, 499OHM 1/2W 1%
HW-VLV	1	V-6865 VALVE	MIXING VALVE
CW-VLV	1	V-6268 VALVE	MIXING VALVE
ZN-T	1	TE-2684P-1	SENS,T-Ni,0.1%,RM

__________________ **1.** The system contains ___ (number) dampers.

2. List each damper designation and size.

____

____

3. Is the outside air (OA) damper normally open or normally closed? Why?

____

____

____________________ **4.** The return air damper is a(n) ___ type of damper.

____________________ **5.** The designation of the heating valve is ___.

____________________ **6.** The description of the HTG-VLV is ___.

____________________ **7.** The designation of the cooling valve is ___.

____________________ **8.** The description of the CLG-VLV is ___.

Section 10.1 Pneumatic Thermostats

REVIEW QUESTIONS

Name ______________________ **Date** ______________

Multiple Choice

______ **1.** A ___ is a sensing device that consists of two different metals joined together.
- A. setpoint
- B. bimetallic element
- C. baseplate
- D. bleedport

______ **2.** A ___ thermostat changes the air pressure to a valve or damper actuator by changing the amount of air that is expelled to the atmosphere.
- A. bleed-type
- B. pilot bleed
- C. single-temperature-setpoint
- D. limit

______ **3.** A ___ thermostat is a pneumatic thermostat that has a third air connection piped to either a manual air regulator or an outside air transmitter.
- A. deadband pneumatic
- B. master/submaster
- C. limit
- D. chilled-water

______ **4.** A winter/summer thermostat is a ___ thermostat that changes the setpoint and action of the thermostat from the winter (heating) to the summer (cooling) mode.
- A. limit
- B. deadband pneumatic
- C. pilot bleed
- D. single-temperature-setpoint

______ **5.** A ___ is a flat piece of metal to which the thermostat components are mounted.
- A. bimetallic element
- B. baseplate
- C. bleedport
- D. nozzle

______ **6.** ___ thermostats are commonly located in the discharge of the heating coil and set at a minimum temperature that protects the coil or system.
- A. Winter/summer
- B. Pneumatic
- C. Direct-acting
- D. Limit

_______________ 7. A ___ pneumatic thermostat is a thermostat that allows no heating or cooling to take place between two temperatures.
- A. pilot bleed
- B. bleed-type
- C. deadband
- D. master/submaster

_______________ 8. The ___ in a pilot bleed thermostat acts like an amplifier.
- A. relay
- B. actuator
- C. internal restrictor
- D. bimetallic element

_______________ 9. During the day, a day/night thermostat setpoint is usually about ___°F.
- A. 36
- B. 48
- C. 60
- D. 72

_______________ 10. A ___ uses air volume amplified by a relay to control the temperature in a building space or area.
- A. bleed-type (one-pipe) humidistat
- B. bleed-type (one-pipe) thermostat
- C. pilot bleed (two-pipe) humidistat
- D. pilot bleed (two-pipe) thermostat

_______________ 11. Because a winter/summer thermostat valve does not change its operation between winter and summer, the ___ must change to accommodate the different medium.
- A. action of the thermostat
- B. setpoint of the thermostat
- C. metallic element
- D. changeover relay

Completion

_______________ 1. A(n) ___ thermostat uses changes in compressed air to control the temperature in individual rooms inside a commercial building.

_______________ 2. ___ pressure is the pressure in the air line that is piped from the thermostat to the controlled device.

_______________ 3. A(n) ___ application is an application in which the thermostat is located in a return air slot that is integral to a light fixture.

_______________ 4. A(n) ___ is a fixed orifice that meters airflow through a port and allows fine output pressure adjustments and precise circuit control.

_______________ 5. ___ is the desired value to be maintained by a system.

__________________ **6.** A(n) ___ is used to override a thermostat to the day temperature during the night mode.

__________________ **7.** A(n) ___ is the range between two temperatures in which no heating or cooling takes place.

__________________ **8.** A(n) ___ is an orifice that allows a small volume of air to be expelled to the atmosphere.

__________________ **9.** A(n) ___ thermostat is a pneumatic thermostat that maintains a temperature above or below an adjustable setpoint.

__________________ **10.** A(n) ___ thermostat increases the branch line pressure as the building space temperature increases and decreases the branch line pressure as the building space temperature decreases.

__________________ **11.** A(n) ___ thermostat is a pilot bleed thermostat that has one setpoint year-round for a building space or area.

__________________ **12.** A(n) ___ relay is a relay that causes the operation of the thermostat to change between two or more modes such as day/night.

__________________ **13.** Day/night thermostats have ___ bimetallic elements.

__________________ **14.** A dual-duct air handling unit supplies both hot and cold air to a sheet metal ___.

__________________ **15.** One use for a bleed-type thermostat is a(n) ___ application.

__________________ **16.** ___ are superior to bleed-type thermostats because of their higher main air capacity.

Winter/Summer Thermostats

__________________ **1.** control line test gauge

__________________ **2.** direct-acting strip

__________________ **3.** sensitivity slider

__________________ **4.** reverse-acting strip

__________________ **5.** heating adjusting screw

__________________ **6.** cooling adjusting screw

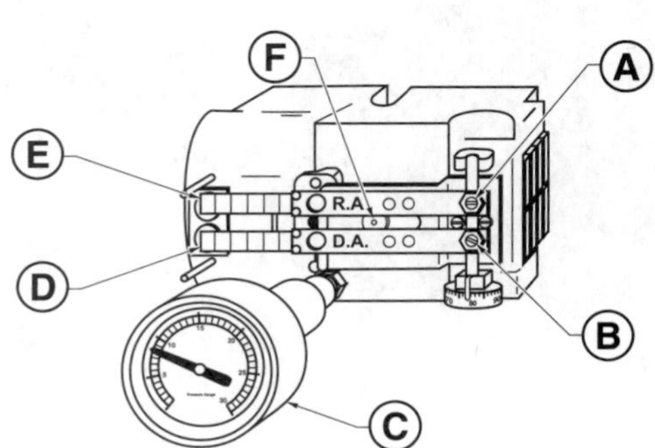

Day/Night Thermostat Calibration

Identify the procedures shown.

______________________ **1.** Adjust night setpoint dial to desired night temperature.

______________________ **2.** Adjust day output to midpoint of actuator spring range.

______________________ **3.** Adjust both setpoint dials to ambient temperature.

______________________ **4.** Change main air pressure to night pressure.

______________________ **5.** Adjust day setpoint dial to desired day temperature.

______________________ **6.** Change main air pressure to day pressure.

______________________ **7.** Remove cover and install pressure gauge.

______________________ **8.** Adjust night output to midpoint of actuator spring range.

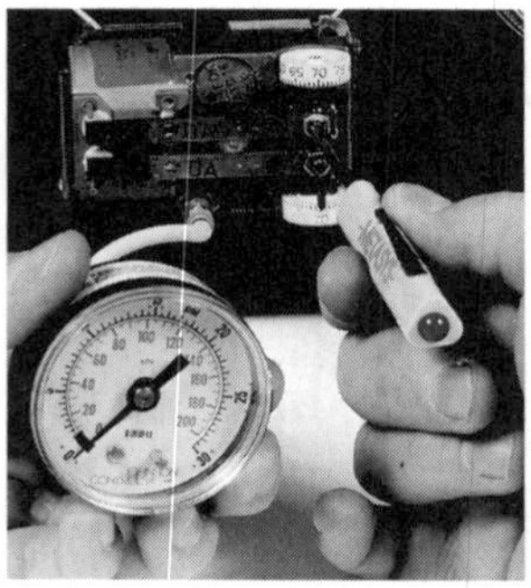

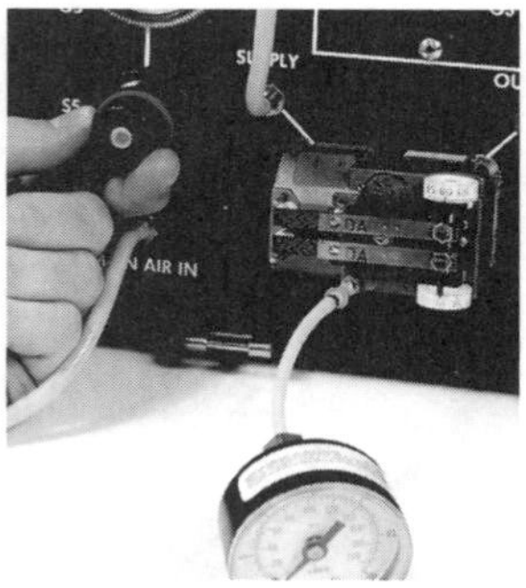

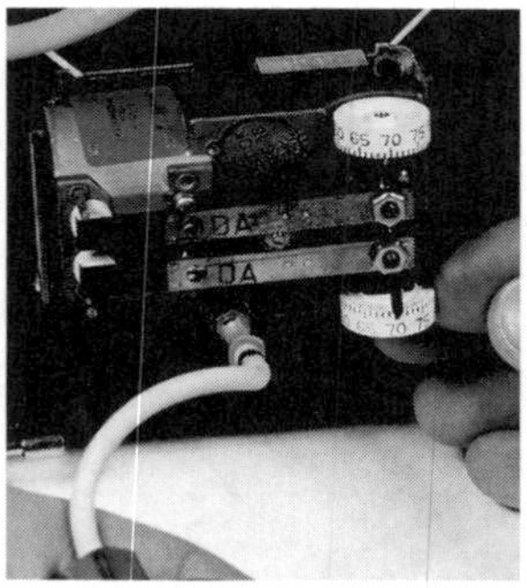

Ⓐ Ⓑ Ⓒ Ⓓ

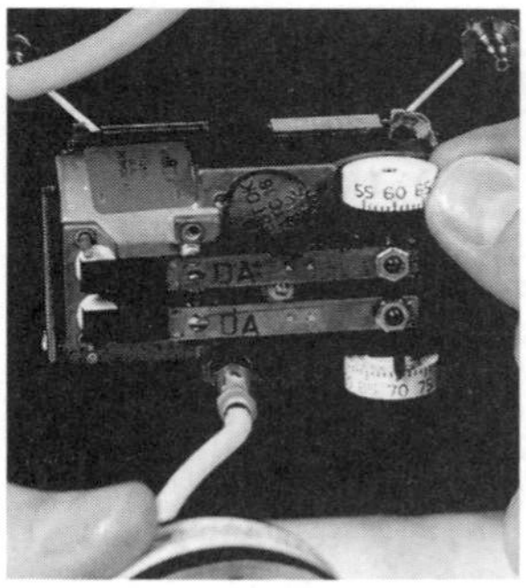

Ⓔ Ⓕ Ⓖ Ⓗ

Limit Thermostats

______________________ **1.** electrical/pneumatic switch

______________________ **2.** limit thermostat

______________________ **3.** pressure switch

______________________ **4.** fan

______________________ **5.** actuator

______________________ **6.** room thermostat

______________________ **7.** N/O steam valve

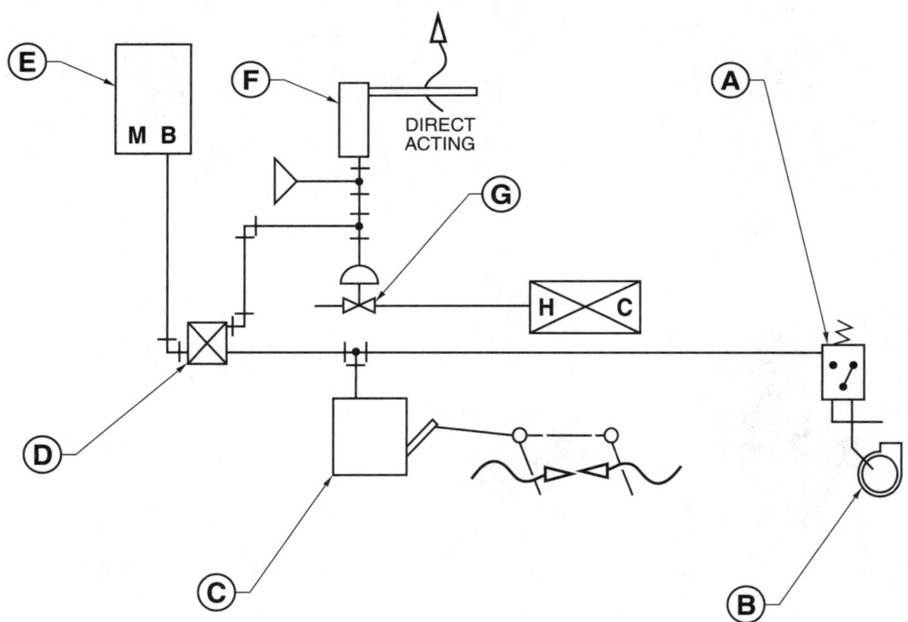

Pilot Bleed Single-Temperature Thermostat Calibration

Identify the procedures shown.

______________ **1.** Adjust branch line pressure to midpoint of actuator spring range.

______________ **2.** Remove pressure gauge and thermostat. Reinstall cover.

______________ **3.** Remove thermostat cover.

______________ **4.** Install pressure gauge and thermometer.

______________ **5.** Check repeatability by turning setpoint dial ±2°F.

______________ **6.** Turn setpoint dial to ambient temperature.

______________ **7.** Turn setpoint dial to desired setpoint.

______________ **8.** Wait 5 min. Measure air temperature near thermostat.

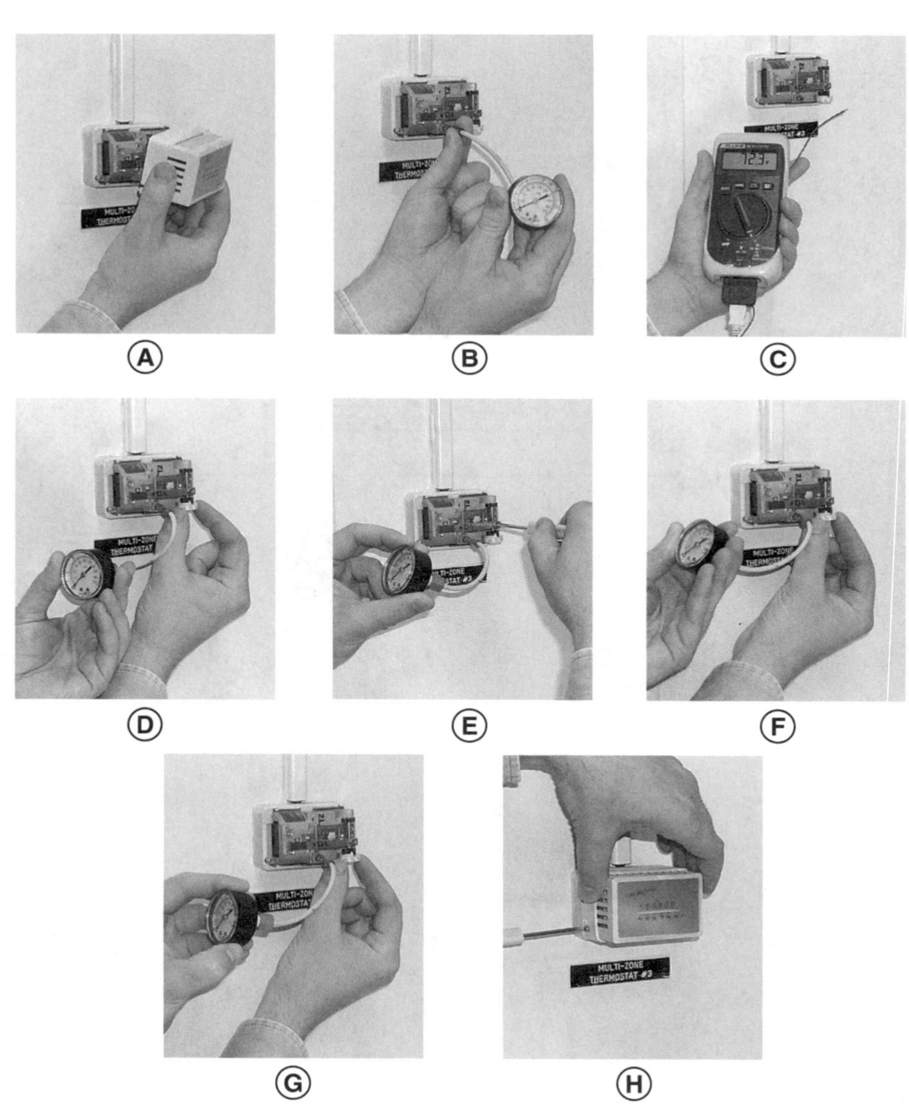

Section 10.2 Pneumatic Humidistats

REVIEW QUESTIONS

Name ______________________________ Date ____________________

Multiple Choice

______________ **1.** The ___ element in a pneumatic humidistat accepts moisture from, or rejects moisture to, the surrounding air.
- A. hygroscopic
- B. hydropic
- C. butyrate
- D. diaphragm

______________ **2.** The construction of a pneumatic humidistat is similar to that of a thermostat, with the exception that the bimetallic element is replaced by a ___ element.
- A. tin alloy
- B. brittle plastic
- C. thin neoprene
- D. hygroscopic

______________ **3.** Some air handling units in applications such as hospitals and manufacturing plants circulate ___ along with the air that can corrode and damage the sensing element.
- A. gases
- B. dirt
- C. water
- D. rust

______________ **4.** Humidistat calibration should be checked at least ___.
- A. semiannually
- B. once a year
- C. twice a year
- D. every five years

______________ **5.** The steam valve is controlled by a ___ humidistat.
- A. reverse-acting single-pipe
- B. direct-acting single-pipe
- C. reverse-acting two-pipe
- D. direct-acting two-pipe

______________ **6.** Calibration of humidity limit controllers is the same as for temperature limit controllers, except that the ambient ___ must be measured accurately.
- A. temperature
- B. humidity
- C. pressure
- D. air balance

Completion

____________________ **1.** A(n) ___ is a controller that uses compressed air to open or close a device which maintains a certain humidity level inside a duct or area.

____________________ **2.** Pneumatic ___ are available in both one-pipe and two-pipe versions.

____________________ **3.** Bleedports are calibrated in exactly the same method as a standard single-temperature ___ thermostat.

____________________ **4.** The steam valve in a humidifier is controlled by a(n) ___ two-pipe humidistat.

____________________ **5.** Almost all humidifier valves are ___ in order to fail the humidifier closed.

____________________ **6.** Humidistat ___ should be checked at least once a year.

Section 10.3 Pneumatic Pressure Switches

REVIEW QUESTIONS

Name ______________________________ **Date** ______________________

Multiple Choice

_______________ **1.** Pneumatic pressure switches are designed to sense and control the pressure inside that duct or building area, which is measured in ___.

A. inches of mercury (Hg)
B. pounds per square inch (psi)
C. inches of water column (in. wc)
D. standard cubic inches of air per minute (scim)

_______________ **2.** Pressure switches can be used to maintain ___ and ensure that the internal building pressure is maintained at an acceptable level.

A. duct air balance parameters
B. internal building humidity
C. indoor air quality
D. indoor air temperature

_______________ **3.** The sensing element of a pneumatic pressure switch is normally a(n) ___ that is connected to the duct or room.

A. aluminum end cap
B. thin neoprene diaphragm
C. copper tube
D. plastic tube

_______________ **4.** Sophisticated pressure control is normally accomplished within ___.

A. dual actuators
B. changeover relays and day/auto levers
C. pressure switches and diaphragms
D. receiver controllers and transmitters

_______________ **5.** Pressure switch calibration should be checked ___.

A. monthly
B. quarterly
C. semiannually
D. annually

Completion

______________________ **1.** A(n) ___ is a controller that maintains a constant air pressure in a duct or area.

______________________ **2.** Specific air pressure is used to supply ___, which open and close to maintain the temperature in the building space.

______________________ **3.** ___ are available in both bleed and pilot bleed (one- and two-pipe) as well as in direct-acting and reverse-acting models.

______________________ **4.** A(n) ___ is piped into the pressure switch by a squeeze bulb or accurate pressure regulator.

ACTIVITIES

Name ______________________ **Date** ______________

Activity 10-1. Troubleshooting VAV Cooling with Finned-Tube Hot Water Heating System

The occupant in a room has called indicating that the room temperature is too hot. It is the first warm spell of the year and the cooling equipment has just started running. The temperature in the room is 80°F. The setpoint of the thermostat is 74°F. Use the drawing to answer the questions.

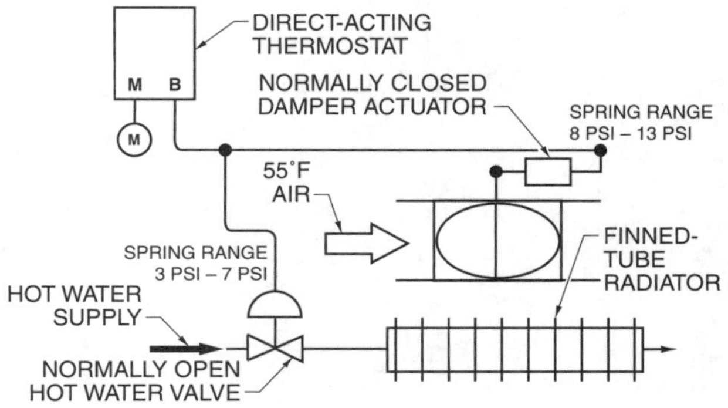

__________ **1.** Is the thermostat in the diagram direct- or reverse-acting?

__________ **2.** As the room temperature increases, should the output pressure increase or decrease?

The cover of the thermostat is removed and an output pressure gauge installed. The output pressure gauge reads 20 psig.

3. What are three possible problems?

____

____

The output pressure can be adjusted from 0 psig to 20 psig easily. As the output pressure is adjusted lower, the supply of 55°F air from the duct seems to decrease and water noise increase, indicating the fin tube coil is heating up.

4. Is the damper or valve stuck? Why or why not?

____

5. If the mechanical system is working, should the calibration of the thermostat be checked? If so, give the procedure.

____

____

Activity 10-2. Troubleshooting Finned-Tube Steam Heating System

The occupant in a room has called indicating that the room temperature was fine until an hour ago but is now too hot. The temperature in the room is 80°F. The setpoint of the thermostat is 72°F. Use the drawing to answer the questions.

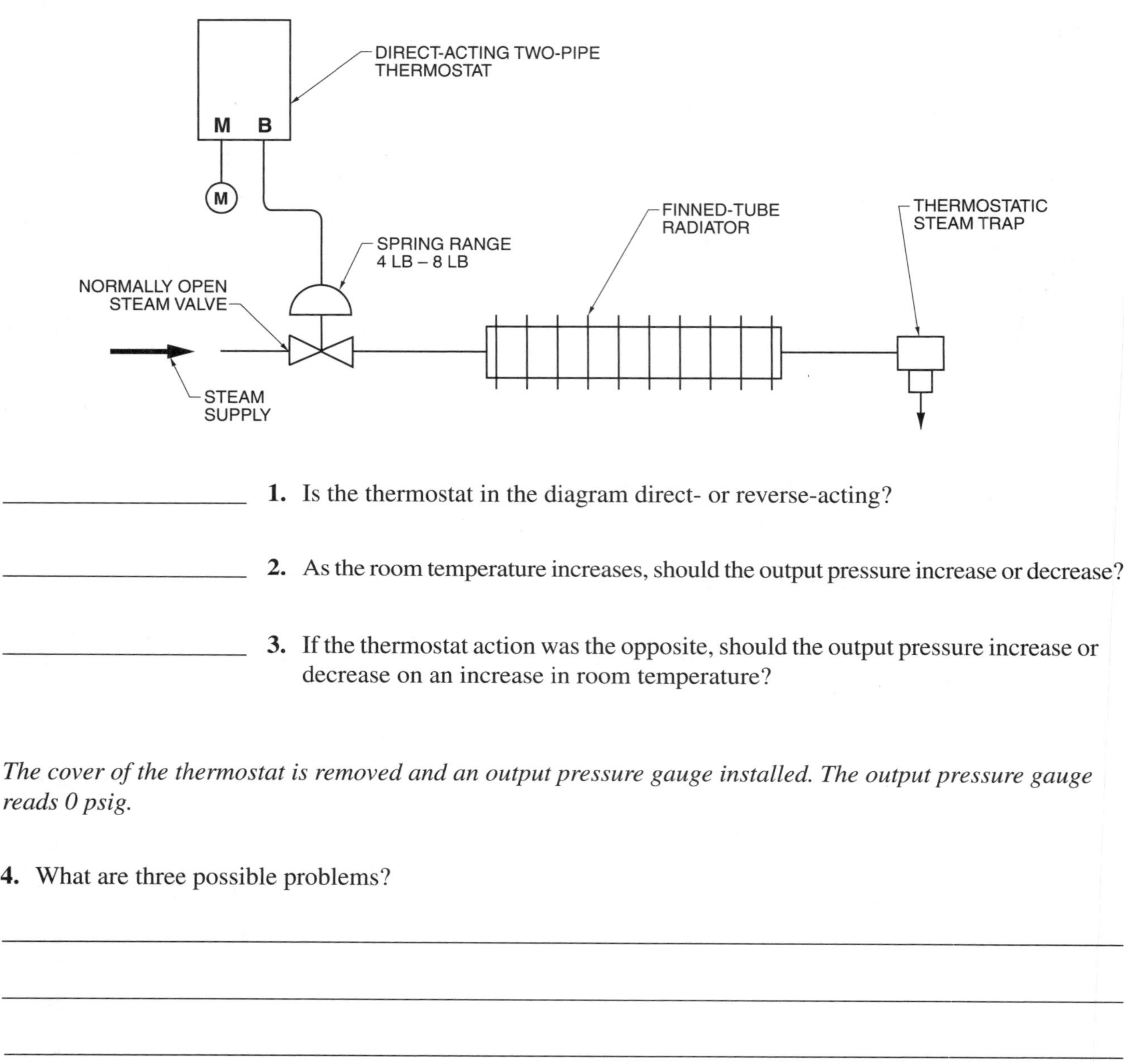

____________________ **1.** Is the thermostat in the diagram direct- or reverse-acting?

____________________ **2.** As the room temperature increases, should the output pressure increase or decrease?

____________________ **3.** If the thermostat action was the opposite, should the output pressure increase or decrease on an increase in room temperature?

The cover of the thermostat is removed and an output pressure gauge installed. The output pressure gauge reads 0 psig.

4. What are three possible problems?

____

____

____

Regardless of the amount the output pressure is adjusted, it remains at 0 psig.

____________________ **5.** What tool is needed to check the operation of the actuator?

The access door to the finned-tube radiator is opened and steam at full flow is observed through the valve. A squeeze bulb is connected to the actuator by removing the air line from the thermostat. An attempt is made to pump air into the actuator and close it, but regardless of the amount of pumping, the actuator pressure does not increase.

6. What is the problem?

____

____

7. Can it be fixed? If so, give the procedure.

____

____

8. How can the actuator be tested to ensure proper operation and no callbacks?

____

____

Section 11.1 Pneumatic Transmitter Characteristics

REVIEW QUESTIONS

Name ______________________________ **Date** ____________________

Multiple Choice

________________ **1.** Transmitter ___ is the output pressure change that occurs per unit of measured variable change.

A. sensitivity
B. range
C. span
D. gauge pressure

________________ **2.** Pneumatic transmitters provide the ___ for HVAC air handling systems, boilers, chillers, and cooling towers.

A. temperature control
B. regulatory function
C. sensing function
D. system control

________________ **3.** ___ transmitters are used in applications that require long piping runs, and they are rarely used in standard building HVAC applications.

A. One-pipe
B. Two-pipe
C. Low-volume
D. High-pressure

________________ **4.** Transmitter ___ is the difference between the minimum and maximum sensing capability of a transmitter.

A. scale
B. gauge
C. range
D. span

________________ **5.** Transmitter sensitivity is found by ___.

A. dividing 12 psig by the transmitter span
B. dividing 12 psig by the transmitter range
C. dividing 15 psig by the transmitter span
D. dividing 15 psig by the transmitter range

________________ **6.** Transmitter receiver gauges are ___ gauges that are marked with the range of the appropriate transmitter.

A. 0 psig to 10 psig
B. 3 psig to 12 psig
C. 3 psig to 15 psig
D. 5 psig to 15 psig

Completion

__________ 1. A(n) ___ is a device that senses temperature, pressure, or humidity and sends a proportional (3 psig to 15 psig) signal to a controller.

__________ 2. A(n) ___ device uses a small amount of the compressed air supply (restricted main air).

__________ 3. A(n) ___ device uses the full volume of compressed air available.

__________ 4. All pneumatic transmitters have a range and ___.

__________ 5. All pneumatic transmitters have a(n) ___ output.

__________ 6. Transmitter ___ is the temperatures between which a transmitter is capable of sensing.

Pneumatic Transmitters

__________ 1. coil

__________ 2. pump

__________ 3. sun shield

__________ 4. reset controller

__________ 5. boiler

__________ 6. hot water transmitter

__________ 7. three-way mixing valve

__________ 8. outside air pneumatic transmitter

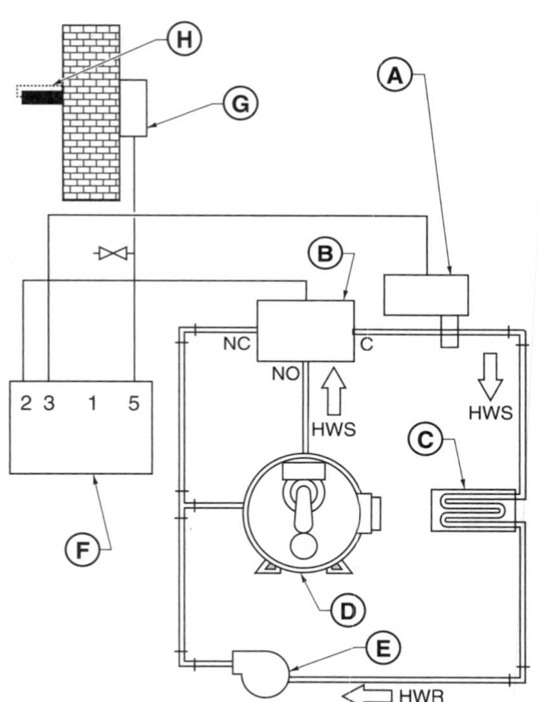

Section 11.2 Temperature Transmitters

REVIEW QUESTIONS

Name ______________________________ **Date** ____________________

Multiple Choice

_______________ **1.** A(n) ___ temperature transmitter is a pneumatic transmitter that uses a long tube filled with a liquid or gas to sense duct temperature.
- A. rod-and-tube
- B. inert-element
- C. averaging element
- D. bulb-type

_______________ **2.** Room temperature transmitters use a(n) ___ for sensing.
- A. setpoint dial
- B. bimetallic strip
- C. strain gauge
- D. output pressure adjustment

_______________ **3.** A(n) ___ temperature transmitter is a pneumatic transmitter that uses a high-quality metal rod with precision expansion and contraction characteristics as the sensing element.
- A. bulb-type
- B. inert-element
- C. room
- D. rod-and-tube

_______________ **4.** A(n) ___ transmitter has a long-rod transmitter section that is connected to a section of metal that has virtually a zero rate of expansion and contraction.
- A. inert-element
- B. averaging
- C. rod-and-tube
- D. bulb-type

_______________ **5.** An averaging element transmitter should be used if the duct diameter is more than ___.
- A. 10′ long
- B. two times the length of the rod element
- C. four times the length of the rod element
- D. the length of the capillary tube and bulb

_______________ **6.** A(n) ___ temperature transmitter is a pneumatic transmitter that uses a capillary tube and bulb filled with a liquid or gas to sense temperature.
- A. rod-and-tube
- B. bulb-type
- C. inert-element
- D. room

Completion

__________ **1.** A(n) ___ transmitter is a pneumatic transmitter that is used to sense outside air temperatures in through-the-wall applications.

__________ **2.** A(n) ___ is a spring or thin piece of metal that measures the movement of a sensing element.

__________ **3.** ___ transmitters have sensing elements 6″ to 12″ long and are used to sense air temperature in air handling systems.

__________ **4.** ___ transmitters have sensing elements approximately 4″ long and are used to sense water temperatures in hot water, chilled water, or cooling tower systems.

__________ **5.** A(n) ___ transmitter is a transmitter used in applications that require a receiver controller to measure and control the temperature in an area.

Rod-and-Tube Temperature Transmitters

__________ **1.** brass tube

__________ **2.** bleed port

__________ **3.** metal rod

__________ **4.** sensor line port

__________ **5.** spring

__________ **6.** flapper/nozzle assembly

Inert-Element Temperature Transmitters

__________ **1.** sensing element shield

__________ **2.** inert section

__________ **3.** inert-element temperature transmitter

__________ **4.** long-rod section (sensing element)

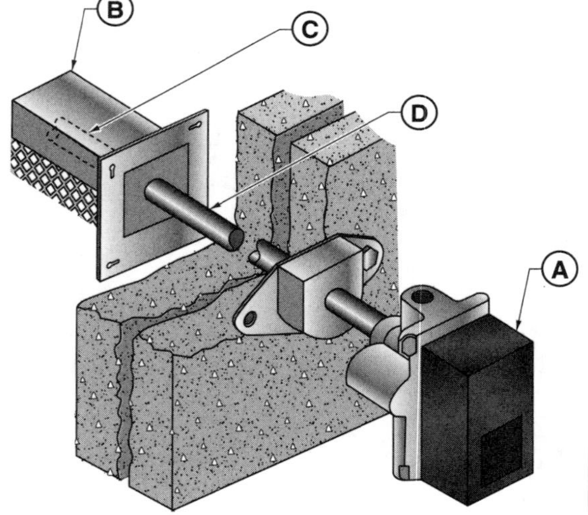

Section 11.3 Humidity Transmitters

REVIEW QUESTIONS

Name ______________________ **Date** ______________

Multiple Choice

_______________ **1.** A ___ transmitter is a device that measures the amount of moisture in the air compared to the amount of moisture the air could hold if it were saturated.

A. room temperature
B. bulb-type
C. duct flow
D. pneumatic humidity

_______________ **2.** A pneumatic humidity transmitter is a device that measures the ___.

A. amount of moisture in the air compared to the amount of moisture the air could hold if it were saturated
B. amount of moisture in the air compared to the amount of moisture the air holds on average
C. average humidity in a duct
D. amount of moisture in the air surrounding the transmitter

_______________ **3.** Humidity transmitters are affected by ___, corrosion, and dirt.

A. moisture
B. drift
C. air pressure
D. temperature

_______________ **4.** Room humidity transmitters are used to sense the relative humidity in a building space (room) and send a ___ signal to a receiver controller.

A. 1 psig to 5 psig
B. 2 psig to 10 psig
C. 3 psig to 15 psig
D. 4 psig to 20 psig

_______________ **5.** The ___ changes size as it absorbs moisture from the air or releases moisture to the air.

A. pitot tube
B. hygroscopic element
C. humidifer valve
D. duct sampling tube kit

Completion

_______________ 1. ___ transmitters are used to sense the relative humidity in a building space and send a 3 psig to 15 psig signal to a receiver controller.

_______________ 2. A(n) ___ allows the air from the air stream to be forced across the humidity-sensing element.

_______________ 3. Humidity transmitters are inaccurate with humidity extremes such as 0% to ___% rh.

Section 11.4 Pressure Transmitters

REVIEW QUESTIONS

Name ______________________________ **Date** ______________________

Multiple Choice

_______________ **1.** A(n) ___ transmitter is a device mounted in a duct that senses the static pressure due to air movement.
   A. duct pressure
   B. pipe pressure
   C. humidity
   D. inert-element

_______________ **2.** Pressure transmitters use sensing elements made of ___ for air.
   A. metal
   B. hard plastic
   C. rubber
   D. neoprene

_______________ **3.** In a pressure transmitter, the sensing element is connected to a(n) ___ through a strain gauge connection that causes the pressure in the air line to vary between 3 psig and 15 psig.
   A. pitot tube
   B. flapper/nozzle assembly
   C. averaging element
   D. capillary tube

_______________ **4.** Many variable-air-volume (VAV) control systems place the transmitter ___ after the fan.
   A. one-third of the distance of the shortest run of ductwork
   B. two-thirds of the distance of the shortest run of ductwork
   C. one-third of the distance of the longest run of ductwork
   D. two-thirds of the distance of the longest run of ductwork

_______________ **5.** Pipe pressure transmitters send a 3 psig to 15 psig pressure to a ___ that performs system control.
   A. damper actuator
   B. reference pressure probe
   C. receiver controller
   D. sensing element shield

Completion

______________________ **1.** ___ transmitters are used in HVAC piping systems to sense the pressure in water distribution systems.

______________________ **2.** A(n) ___ is a device that senses static pressure and total pressure in a duct.

______________________ **3.** A(n) ___ transmitter is a device used to sense the pressure due to airflow in a duct or water flow through a pipe.

Duct Pressure Transmitters

______________________ **1.** pitot tube

______________________ **2.** discharge air

______________________ **3.** normally open damper

______________________ **4.** receiver controller

______________________ **5.** duct pressure transmitter

______________________ **6.** duct

______________________ **7.** static pressure

______________________ **8.** pneumatic damper actuator

______________________ **9.** static pressure probe

______________________ **10.** reference pressure probe

______________________ **11.** total pressure

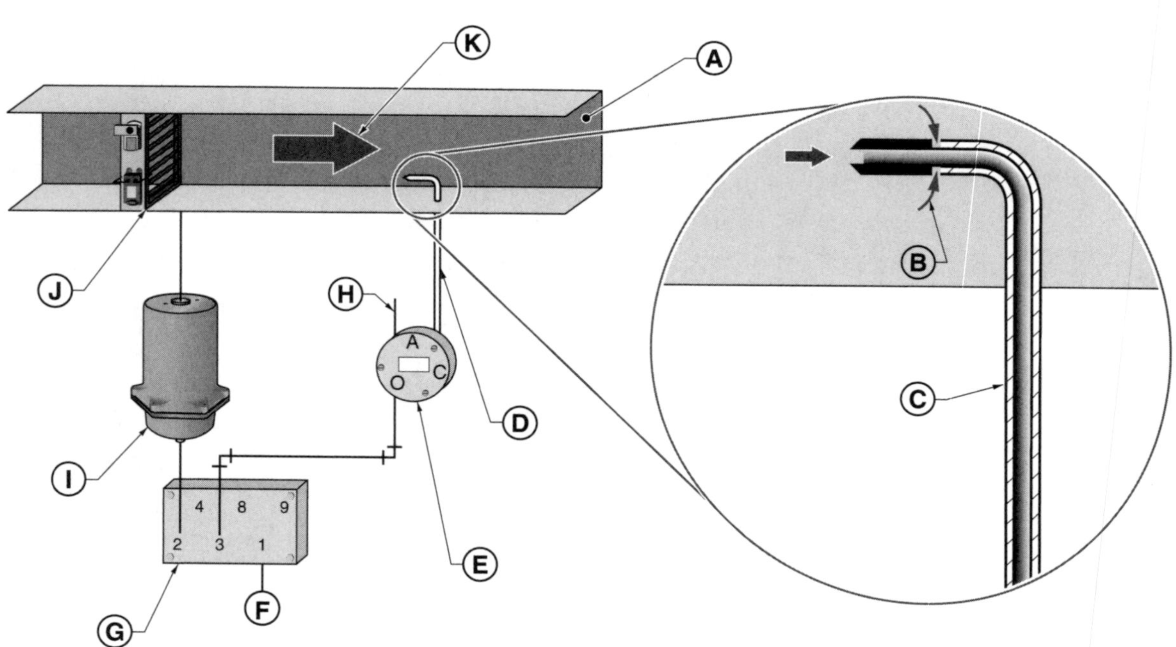

Pneumatic Transmitters

ACTIVITIES

Name ______________________________ **Date** ______________

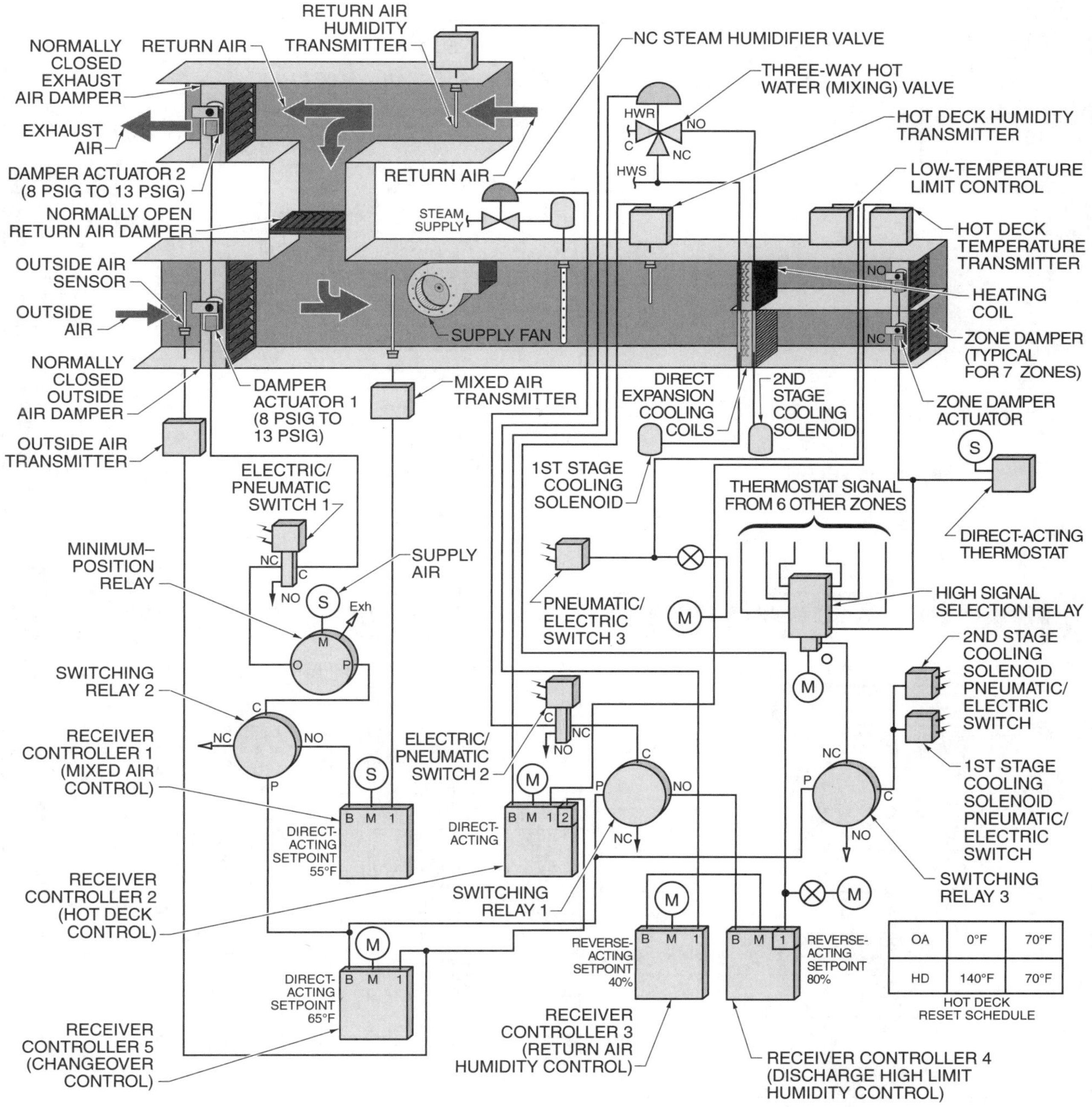

OA	0°F	70°F
HD	140°F	70°F

HOT DECK RESET SCHEDULE

TRANSMITTER CHARACTERISTICS

Temperature Range*	Span*	Sensitivity†
40 – 65 60 – 85	25	0.48
50 – 100	50	0.24
0 – 100 20 – 120 50 – 150	100	0.12
40 – 240 –40 – 160 200 – 400	200	0.06

* in °F
† in psi/°F

TEMPERATURE TRANSMITTER TYPES

Order No.	Sensing Range (Non-Adjustable)*	Element Length	Mounting
MA825B 1011	–40 to 160	27‡	Wall
MA825B 1029	40 to 240	15″	Well
MA825B 1037	–40 to 160	15″	Well
MA825B 1052	40 to 240	7″	Well
MA825B 1060	–40 to 160	7″	Well
MA825B 1110	–20 to 80	15″	Well
MA825B 1250	–20 to 80	27‡	Wall
Order No.	**Sensing Range (Non-Adjustable)***	**Element Length**	**Mounting**
MA825B 1003	–40 to 160	15″	Duct
MA826B 1044	0 to 200	8 ⅞′	Duct
MA825B 1045	–40 to 160	7″	Duct
MA826B 1051	0 to 200	8 ⅞′	Duct
MA826B 1077	25 to 125	18 ½′	Duct
MA826B 1085	25 to 125	8 ⅞′	Duct
MA825B 1243	–20 to 80	15″	Duct
MA825B 1268	40 to 240	15″	Duct

* in °F
‡ active element 15″, inert section 12″

TEMPERATURE AND RELATIVE HUMIDITY vs. TRANSMITTER OUTPUT PRESSURE

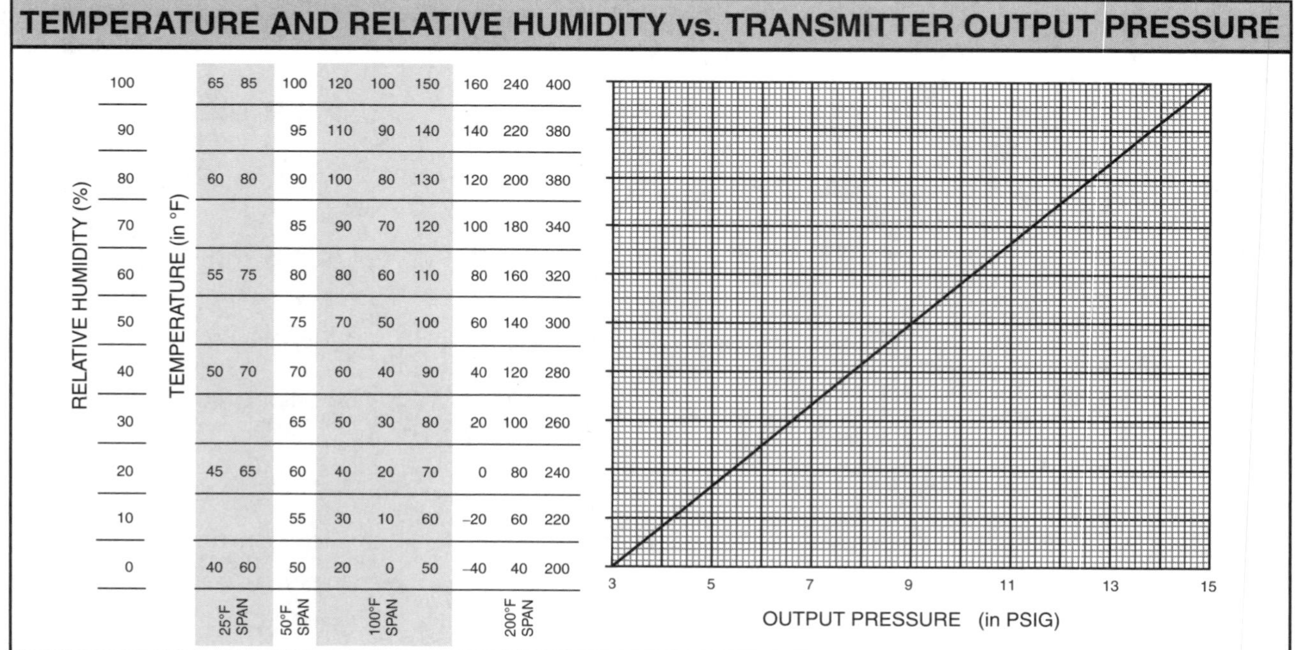

Activity 11-1. Mixed Air Pneumatic Transmitter Selection

Locate the mixed air transmitter on the multizone air handling unit drawing and label it T-1 for your reference. The mixed air transmitter is installed inside a large air duct and must obtain an accurate air temperature sample. The expected value of the mixed air temperature is approximately 55°F.

1. Using the Temperature Transmitter Types chart, what possible transmitter models may be selected?

____

____

2. What factors determine the selection?

____

____

______________ **3.** The most appropriate transmitter temperature range is ___°F.

Use the Transmitter Characteristics chart to determine the transmitter span and sensitivity.

______________ **4.** The selected transmitter span is ___°F.

______________ **5.** The selected transmitter sensitivity is ___ psig/°F.

Use the Temperature and Relative Humidity vs. Transmitter Output Pressure chart to determine the output pressure.

______________ **6.** The output pressure of the selected transmitter at 60°F and 40% rh is ___ psig.

______________ **7.** The output pressure of the selected transmitter at 40°F and 20% rh is ___ psig.

______________ **8.** The output pressure of the selected transmitter at 80°F and 30% rh is ___ psig.

Activity 11-2. Hot Deck Pneumatic Transmitter Selection

Locate the hot deck transmitter on the multizone air handling unit drawing and label it T-2 for your reference. The hot deck transmitter is installed inside a small air duct.

______________ **1.** Using the hot deck reset schedule, what is the expected hot deck temperature range?

2. Using the Temperature Transmitter Types chart, what possible transmitter models may be selected?

____

____

3. What factors determine the selection?

____

____

________________ **4.** The most appropriate transmitter temperature range is ___°F.

Use the Transmitter Characteristics chart to determine the transmitter span and sensitivity.

________________ **5.** The selected transmitter span is ___°F.

________________ **6.** The selected transmitter sensitivity is ___ psig/°F.

Use the Temperature and Relative Humidity vs. Transmitter Output Pressure chart to determine the output pressure.

________________ **7.** The output pressure of the selected transmitter at 60°F and 50% rh is ___ psig.

________________ **8.** The output pressure of the selected transmitter at 100°F and 30% rh is ___ psig.

________________ **9.** The output pressure of the selected transmitter at 150°F and 55% rh is ___ psig.

Activity 11-3. Outside Air Pneumatic Transmitter Selection

Locate the outside air temperature transmitter on the multizone air handling unit drawing and label it T-3 for your reference. The outside air temperature transmitter is installed through the wall into the outside air.

1. Using the Temperature Transmitter Types chart, what possible transmitter models may be selected?

____

____

2. What factors determine the selection?

____

____

________________ **3.** Based on the local climate, what is the expected outside air temperature range?

________________ **4.** The most appropriate transmitter temperature range is ___°F.

Use the Transmitter Characteristics chart to determine the transmitter span and sensitivity.

________________ **5.** The selected transmitter span is ___°F.

________________ **6.** The selected transmitter sensitivity is ___ psig/°F.

Use the Temperature and Relative Humidity vs. Transmitter Output Pressure chart to determine the output pressure.

______________ **7.** The output pressure of the selected transmitter at 35°F and 35% rh is ___ psig.

______________ **8.** The output pressure of the selected transmitter at 75°F and 55% rh is ___ psig.

______________ **9.** The output pressure of the selected transmitter at 95°F and 45% rh is ___ psig.

Section 12.1 Receiver Controllers and Transmitters

REVIEW QUESTIONS

Name ______________________________ Date ______________

Multiple Choice

______________ **1.** A ___ is a device that consists of a controller mounted to a duct or pipe and that is connected by a capillary tube to a bulb that is inserted into the duct or pipe.

A. single-input receiver controller
B. dual-input receiver controller
C. remote bulb controller
D. force-balance receiver controller

______________ **2.** ___ controllers are normally not used to control the temperature, humidity, and pressure in one room or zone of a commercial building.

A. Fluidic
B. Receiver
C. Remote bulb
D. Force-balance receiver

______________ **3.** A ___ is a device that sends an air pressure signal to a receiver controller.

A. restrictor
B. pneumatic transmitter
C. fluidic controller
D. temperature transmitter

______________ **4.** A ___ is a fixed orifice that meters airflow through a port and allows fine output pressure adjustments and precise circuit control.

A. pneumatic signal limiter
B. valve actuator
C. restrictor
D. transmitter

Completion

______________ **1.** A(n) ___ is a device which accepts one or more input signals from pneumatic transmitters and produces an output signal based on its calibration.

______________ **2.** A(n) ___ design is a design in which the controller output is determined by the relationship of mechanical pressures.

______________ **3.** A pneumatic ___ senses temperature, pressure, or humidity and sends a proportional 3 psig to 15 psig signal to a receiver controller.

_______________ 4. A(n) ___ may be required to prevent undue pressure drop in extremely long piping runs (over 250′).

_______________ 5. A(n) ___ uses vector analysis to arrive at an output.

Receiver Controllers

_______________ 1. master test point

_______________ 2. remote or local setpoint test point

_______________ 3. gain adjustment dial

_______________ 4. controlled variable

_______________ 5. readjustment side

_______________ 6. supply

_______________ 7. output

_______________ 8. controlled variable test point

_______________ 9. internal supply

_______________ 10. setpoint dial

_______________ 11. ratio adjustment dial

_______________ 12. master

_______________ 13. ratio selection jumper

_______________ 14. setpoint (remote)

_______________ 15. internal supply test point for factory use only

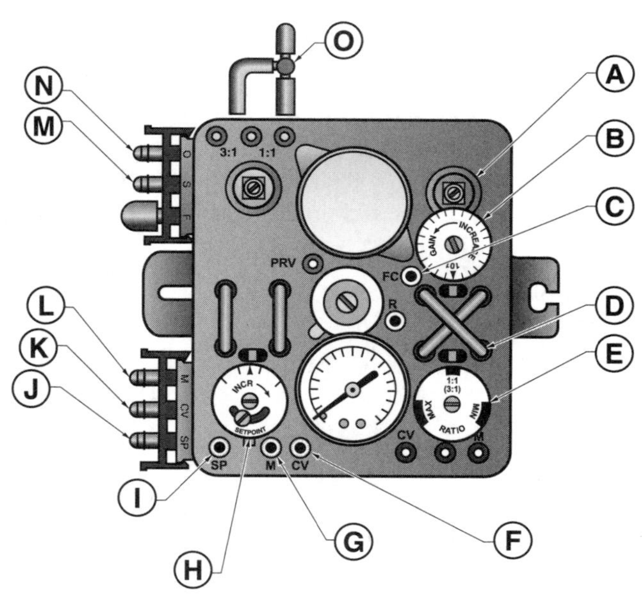

Section 12.2 Single-Input Receiver Controllers

REVIEW QUESTIONS

Name ______________________________ Date ____________________

Multiple Choice

__________ **1.** ___ is the number of units of controlled variable that causes an actuator to move through its entire spring range.

A. Tubing jumper band
B. Point adjustment
C. Proportional band
D. Gain

__________ **2.** ___ air is used in an air handling unit for cooling when the system is in the economizer mode.

A. Return
B. Makeup
C. Outside
D. Mixed

__________ **3.** Calibration of a single-input receiver controller without a ___ yields less precise results.

A. simulator kit
B. receiver controller
C. fluidic controller
D. reset control

Completion

__________ **1.** A(n) ___ controller is designed to be connected to only one transmitter and to maintain only one temperature, pressure, or humidity setpoint.

__________ **2.** ___ is the mathematical relationship between the controller output pressure change and the transmitter pressure change that causes it.

__________ **3.** Single-input receiver controllers are often used to control vital ___ systems in a building.

Hot Water Control

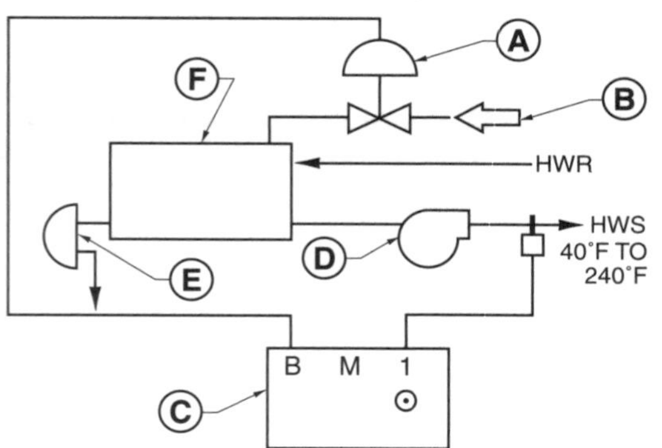

______________ 1. steam trap

______________ 2. steam supply

______________ 3. steam valve

______________ 4. heat exchanger

______________ 5. direct-acting controller

______________ 6. pump

Single-Input Receiver Controller Calibration

Identify the procedures shown.

______________ 1. Adjust output pressure of controller to midpoint of actuator spring range.

______________ 2. Disconnect transmitter air line and connect simulator to transmitter port.

______________ 3. Check throttling range and gain adjustment. Readjust if necessary.

______________ 4. Determine transmitter range, setpoint, throttling range, gain, and actuator spring range.

______________ 5. Enter desired throttling range and gain on controller.

______________ 6. Remove simulator, reattach air lines, and check actual operation.

______________ 7. Adjust simulator pressure-regulating valve to desired setpoint.

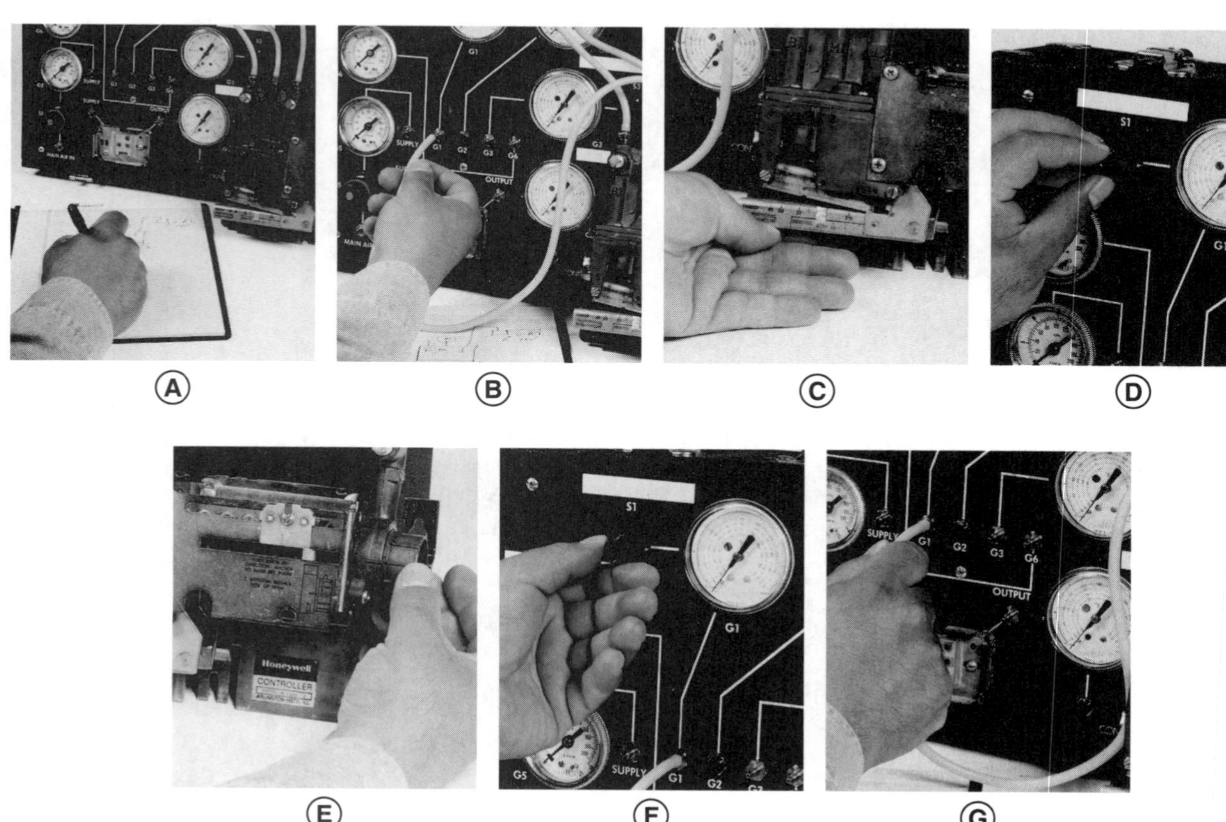

Section 12.3 Dual-Input Receiver Controllers

REVIEW QUESTIONS

Name ______________________________ **Date** ______________________

Multiple Choice

_______________ **1.** Central plant hot water and chilled water systems typically use a ___ controller to save energy and increase comfort by changing setpoints when outside air temperature or load conditions change.

A. force-balance receiver
B. single-input receiver
C. remote bulb
D. dual-input receiver

_______________ **2.** A(n) ___ receiver controller is a receiver controller in which the change of one variable, commonly outside air temperature, causes the setpoint of the controller to automatically change (reset) to match the changing condition.

A. single-input
B. dual-input
C. remote bulb
D. externally restricted

_______________ **3.** ___ is a schedule in which the primary variable increases as the secondary variable decreases, and decreases as the secondary variable increases.

A. Reset schedule
B. Integration schedule
C. Reverse readjustment (winter reset)
D. Direct readjustment (summer reset)

_______________ **4.** A ___ is a chart that describes the setpoint changes in a pneumatic control system.

A. reset schedule
B. vector analysis
C. troubleshooting chart
D. feedback chart

_______________ **5.** ___ is the relationship of the primary variable change to the secondary variable change, expressed as a percentage.

A. Proportional band
B. Throttling range
C. Authority
D. Readjustment

_______________ **6.** Direct acting and reverse acting describe the relationship between the primary variable and the ___.
A. setpoints
B. secondary variable
C. output pressure
D. input pressure

_______________ **7.** All receiver controllers should be recalibrated a minimum of once per ___.
A. day
B. week
C. month
D. year

_______________ **8.** Energy wasted due to overcooling may be solved by automatically changing the amount of cooling delivered based on the ___.
A. actual amount of cooling demanded by the highest zone
B. actual amount of cooling demanded by the lowest zone
C. temperature of the farthest space compared to the set temperature range
D. average temperature of all zones affected by the air handling unit

_______________ **9.** A system set for the worst-case scenario can cause energy waste and a lack of comfort, which can be solved by ___.
A. manually changing the amount of cold air delivered based on the outside air temperature
B. automatically changing the amount of hot air delivered based on the outside air temperature
C. manually resetting the hot deck reset control
D. readjusting the dual-input receiver controller

_______________ **10.** A hot water transmitter with a range of 40°F to 240°F, a proportional band of 5%, and a throttling range of 10% has an authority of ___%.
A. 157
B. 183
C. 201
D. 246

Completion

_______________ **1.** Dual-duct and multizone air handling units have a(n) ___ configuration.

_______________ **2.** As the water temperature increases in a hot water heating system, the controller causes the valve to close, reducing the steam flow into the heat exchanger and causing the water temperature to ___.

_______________ **3.** The most common problem with ___ is the maladjustment of the authority/ratio settings.

Section 12.4 Remote Adjustment and Controller Functions

REVIEW QUESTIONS

Name ______________________ **Date** ______________________

Multiple Choice

_______________ **1.** A receiver controller can be configured to function as a proportional controller or as a ___ controller.
A. fluidic
B. remote bulb
C. dual-input receiver
D. proportional/integral

_______________ **2.** Changing the setpoint of a receiver controller is accomplished by changing the ___.
A. air temperature
B. air pressure
C. controller calibration
D. component arrangement

_______________ **3.** The total amount of remote control point adjustment change allowed is expressed as a percentage of the primary transmitter span, usually ___.
A. 1% or 2%
B. 5% or 8%
C. 10% or 20%
D. 25% or 30%

_______________ **4.** The setpoint ___ as the remote control point adjustment pressure increases.
A. increases
B. remains constant
C. fluctuates
D. decreases

_______________ **5.** In single- or dual-input receiver controllers, the standard calibration procedure should be followed except the controller must be calibrated ___.
A. by resetting the controller remotely
B. manually in the maintenance operator's office
C. with a setpoint of zero
D. with the control point adjustment pressure midpoint connected to the control point adjustment port

______________ **6.** Any controller that has not been using the remote control point adjustment function must be ___ before remote control point adjustment can be used.
   A. recalibrated
   B. replaced
   C. integrated
   D. custom-calibrated

______________ **7.** Controllers using integration should first be calibrated with the integration set at ___.
   A. 0
   B. 2
   C. 5
   D. 10

______________ **8.** A longer integration time causes a system to respond more slowly, with ___.
   A. a higher possibility of hunting
   B. increased cycling
   C. increased stability
   D. decreased stability

Completion

______________ **1.** ___ is the ability to adjust the controller setpoint from a remote location.

______________ **2.** ___ is a function that calculates the amount of difference between the setpoint and control point (offset) over time.

______________ **3.** Controllers that add integration are referred to as ___ controllers.

Pneumatic Receiver Controllers

ACTIVITIES

Name ______________________________ **Date** ______________

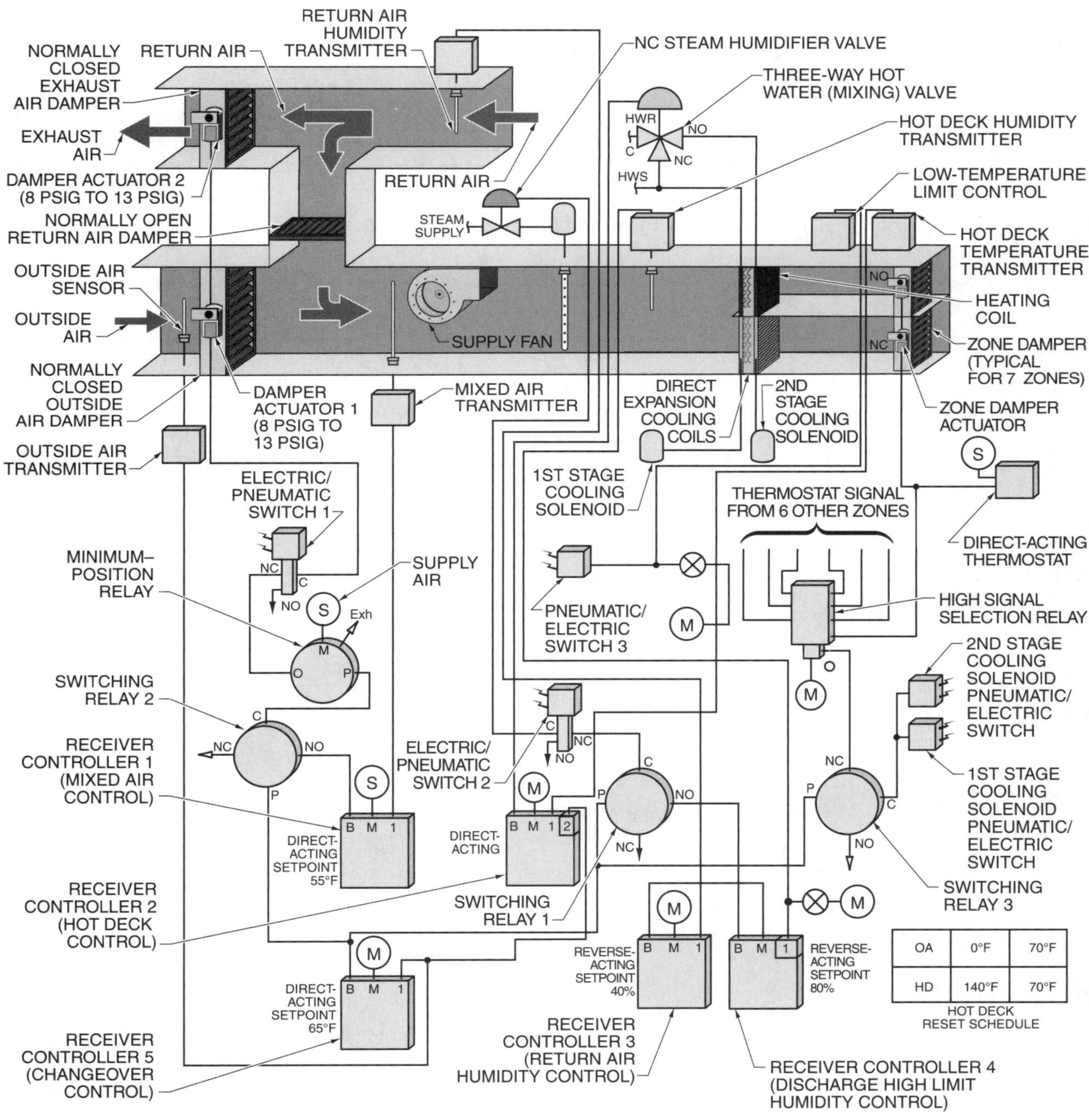

OA	0°F	70°F
HD	140°F	70°F

HOT DECK RESET SCHEDULE

Activity 12-1. Changeover Control

Locate receiver controller 5 (changeover control) on the multizone air handling unit drawing and answer the questions.

__________ **1.** Is receiver controller 5 a single- or dual-input receiver controller?

__________ **2.** The controller setpoint is ___.

__________ **3.** The controller action is ___.

4. List all the devices piped to the branch line of the controller.

____

____

5. Is the controller internally or externally restricted? How can this be checked?

____

____

__________ **6.** At an outside air temperature of 77°F, the output pressure from the controller is ___ psig.

__________ **7.** At an outside air temperature of 35°F, the pressure at port P of switching relay 1 is ___ psig.

Activity 12-2. Hot Deck Control

Locate receiver controller 2 (hot deck control) on the multizone air handling unit drawing and answer the questions.

__________ **1.** Is receiver controller 2 a single- or dual-input receiver controller?

__________ **2.** The controller setpoint/reset schedule is ___.

__________ **3.** At a 35°F outside air temperature, the hot deck setpoint is ___°F.

__________ **4.** The controller action is ___.

__________ **5.** The controller readjustment is ___.

6. List all the devices piped to the branch line of the controller.

____

____

__________ **7.** The ___ transmitter piped to the controller is internally restricted.

Activity 12-3. Humidity Control

Locate receiver controller 3 (return air humidity control) and receiver controller 4 (discharge high limit humidity control) on the multizone air handling unit drawing and answer the questions.

____________________ **1.** Are receiver controllers 3 and 4 single- or dual-input receiver controllers?

____________________ **2.** Receiver controller 3 setpoint is ___.

____________________ **3.** Receiver controller 4 setpoint is ___.

____________________ **4.** Receiver controller 3 action is ___.

____________________ **5.** Receiver controller 4 action is ___.

____________________ **6.** The branch line of receiver controller 3 is piped to ___.

____________________ **7.** At a return air relative humidity of 30%, the output pressure of receiver controller 3 is ___ psig.

8. At a discharge air relative humidity of 85%, is receiver controller 3 able to open the humidifier valve? Why or why not?

____

____

____________________ **9.** At an outside air temperature of 75°F, the pressure at port P of switching relay 1 is ___ psig.

Activity 12-4. Receiver Controller Troubleshooting

Scenario 1: A complaint is received that the air is too hot. The outside air temperature is 75°F. The mixed air dampers are wide open.

____________________ **1.** The output pressure gauge on receiver controller 5 is 0 psig. Is this correct?

2. How should the operation of the controller be checked?

____

____

Scenario 2: A complaint is received that the air is too cold. The outside air temperature is 35°F. Inspection of the unit indicates that the hot duct temperature is approximately 85°F.

____________________ **3.** Is this correct?

4. How should the operation of the controller be checked?

____

____

Scenario 3: A complaint is received that the humidity level is too low. The return air humidity is measured and found to be 30% rh and the discharge air humidity is 27% rh.

____________________ **5.** The output pressure gauge on receiver controller 3 reads 20 psig. Is this correct?

____________________ **6.** The output pressure gauge on receiver controller 4 reads 0 psig. Is this correct?

7. Which controller has a problem and how would it be corrected?

____

____

Section 13.1 Auxillary Devices

REVIEW QUESTIONS

Name ______________________________ **Date** ____________________

Multiple Choice

______________ **1.** A(n) ___ is a device used in a control system that produces a desired function when actuated by the output signal from a controller.
A. controller
B. auxiliary device
C. controlled device
D. booster relay

______________ **2.** A booster relay is a device that ___ the air volume available to a damper or valve while maintaining the air pressure at a 1:1 ratio.
A. increases
B. decreases
C. maintains
D. varies

______________ **3.** A ___ is a multiple-input device that selects the higher or lower of two pneumatic signal levels.
A. controller
B. transducer
C. booster relay
D. signal selection relay

______________ **4.** Biasing relays enable multiple heating and cooling devices to be placed in the ___.
A. same cabinet
B. correct sequence
C. same system
D. proper spring range

______________ **5.** An electronic/pneumatic transducer (EPT) is a device that converts an electronic input signal to a(n) ___ signal.
A. digital output
B. electrical output
C. air pressure output
D. electronic output

______ 6. Low temperature cutout control setpoints are adjustable, with the most common being between 35°F and ___°F.
A. 32
B. 35
C. 40
D. 65

______ 7. A ___ relay is a device that switches airflow from one circuit to another.
A. booster
B. multiple signal selection
C. low signal selection
D. switching

______ 8. A ___ relay is a relay that provides measurement of a large number of zones to ensure accurate zone signal measurement.
A. multiple signal selection
B. switching
C. booster
D. low signal selection

______ 9. ___ switches perform a task that is the opposite of that of electric/pneumatic switches, and the switches cannot be interchanged.
A. Pneumatic/electric
B. Electronic/pneumatic
C. Cutout
D. Biasing

Completion

______ 1. A(n) ___ is a device that allows two different types of components, voltage levels, voltage types, or systems to be interconnected.

______ 2. Switching relays are used to create ___ circuits.

______ 3. The branch line pressure of pneumatic direct-acting thermostats ___ as temperature increases.

______ 4. A(n) ___ is a relay that prevents outside air dampers from completely closing.

______ 5. A(n) ___ is an auxiliary device mounted to a damper or valve actuator that ensures that the damper or actuator moves to a given extension.

______ 6. A(n) ___ is a device that allows an air pressure signal to energize or de-energize an electrical device such as a fan, pump, compressor, or electric heating device.

______ 7. A(n) ___ is a device that changes one type of proportional control signal into another.

______ 8. A(n) ___ is a device that protects against damage due to a low temperature condition.

______ 9. Auxiliary devices that change pressure do not change the ___ of airflow.

______ 10. A(n) ___ is a signal selection relay that selects the higher of two input pressures and outputs the higher pressure to a controlled device.

Switching Relays

______________ 1. valve

______________ 2. low-limit transmitter

______________ 3. restrictor

______________ 4. coil

______________ 5. thermostat

______________ 6. switching relay

______________ 7. main air supply

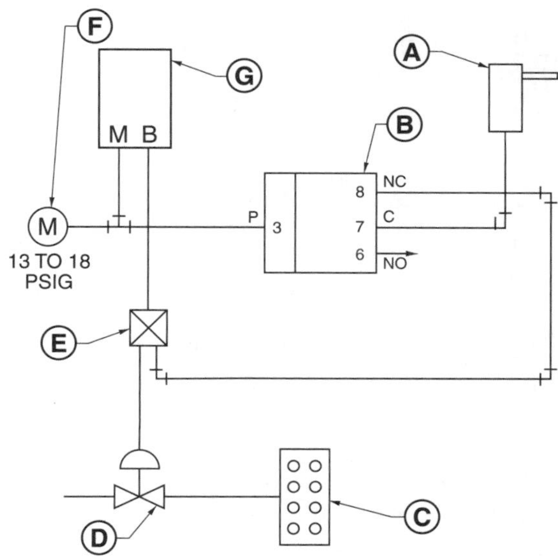

Electronic/Pneumatic Transducers

______________ 1. wiring to electronic circuit

______________ 2. tubing to pneumatic circuit

______________ 3. electronic/pneumatic transducer

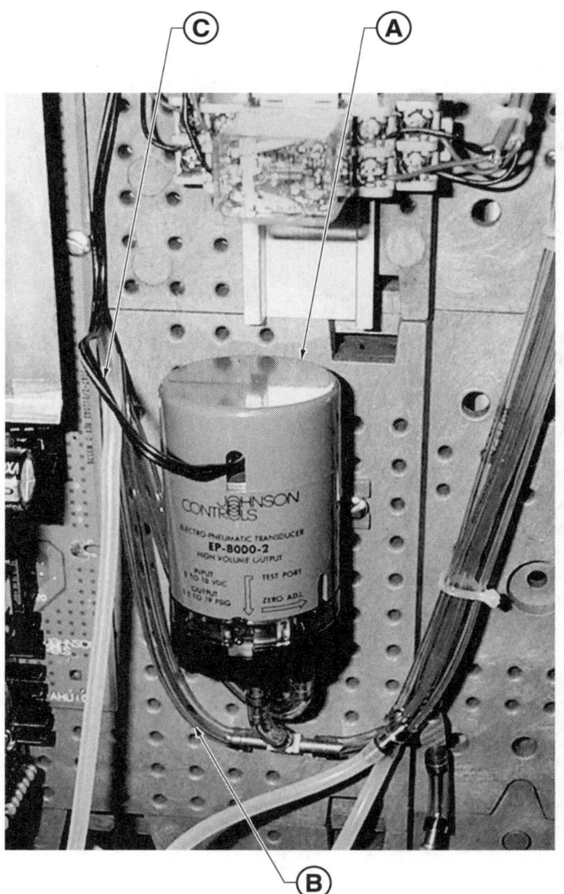

Low Signal Selection Relays

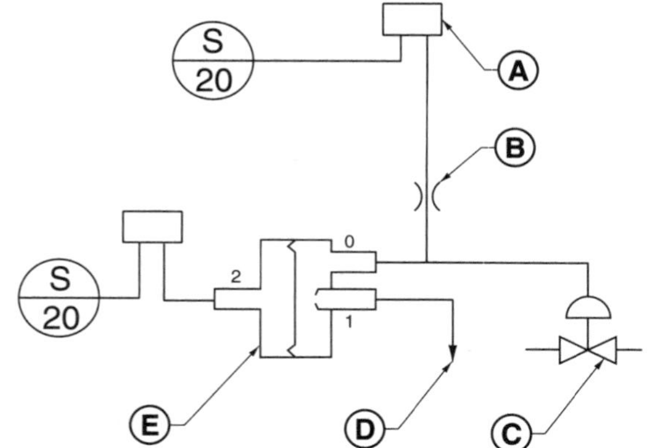

_______________ **1.** exhaust

_______________ **2.** direct-acting pneumatic thermostat

_______________ **3.** low signal selection relay

_______________ **4.** restrictor

_______________ **5.** normally open heating valve

Minimum-Position Relays

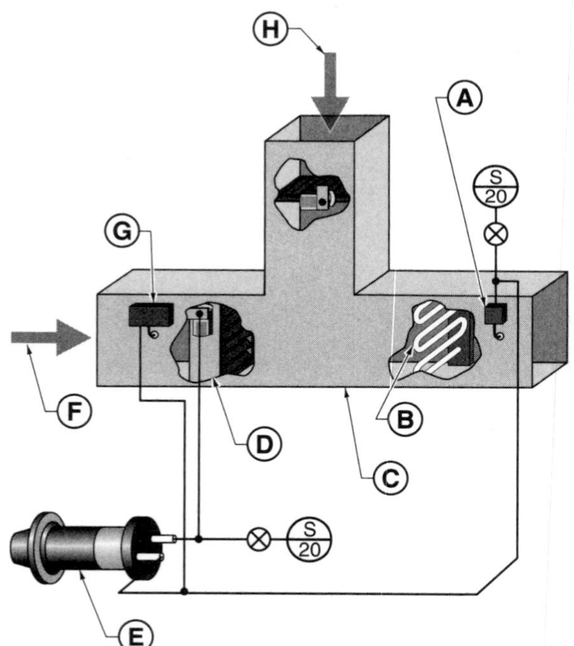

_______________ **1.** mixed-air plenum

_______________ **2.** outside air lockout

_______________ **3.** pneumatic low-limit thermostat

_______________ **4.** cool outside air

_______________ **5.** outside air damper

_______________ **6.** averaging element

_______________ **7.** minimum position relay

_______________ **8.** warm return air

REVIEW QUESTIONS

Name ______________________ **Date** ______________

Multiple Choice

______ **1.** ___ is a common problem associated with auxiliary devices.

A. Equipment operation lockout
B. Unreliable network connections
C. Device connection incompatibility
D. Equipment burnout

______ **2.** When the switching relay does not allow the outside air dampers to go to their minimum position, allowing excessive hot, humid air to enter into the building space, the problem is caused by ___.

A. incorrect input and output pressures from the switching relay
B. a lack of heat caused by a bad or incorrectly calibrated or connected pneumatic/electric switch
C. the normally open port and common port being connected instead of the normally closed port and common port
D. an improper sequence

______ **3.** A ___ is connected to the air pressure input to check and calibrate the pneumatic/electric switch.

A. pressure gauge
B. squeeze bulb
C. calibration kit
D. thermostat

______ **4.** During summer cooling season, fully open outside air dampers can ___ and causr it to be too hot in the building space.

A. bypass the cooling coil
B. bypass the heating coil
C. overload the cooling coil
D. overload the heating coil

______ **5.** The input and output pressures of a minimum-position relays can be checked using ___.

A. pressure guages
B. squeeze bulbs
C. calibration kits
D. bypass calibrators

Completion

____________________ **1.** Thermostats in signal selection systems must be calibrated at the same output pressure and have ___ setpoints.

____________________ **2.** An electric/pneumatic solenoid that is burned out prevents main air from passing through to the ___.

____________________ **3.** ___ can be used to prevent occupants from tampering with the thermostats.

Pneumatic Auxiliary Devices

ACTIVITIES

Name ______________________________ **Date** ________________

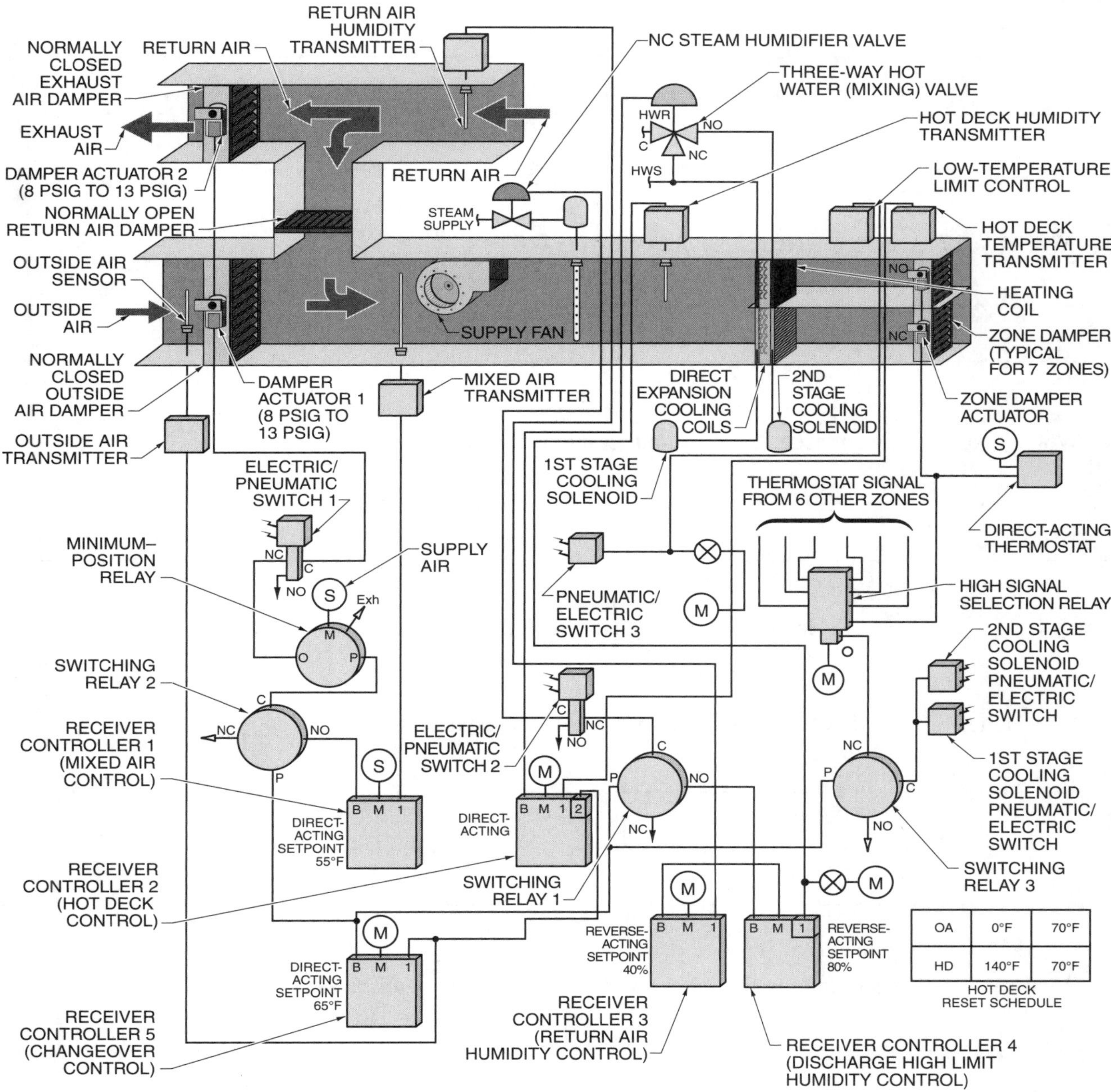

OA	0°F	70°F
HD	140°F	70°F

HOT DECK RESET SCHEDULE

Activity 13-1. Switching Relays

Locate all switching relays on the multizone air handling unit drawing and answer the questions.

____________________ **1.** How many switching relays are used in the multizone air handling unit?

2. List the switching relays.

____

____

____

3. Which two ports are connected if the pilot pressure is 0 psig?

____

____

4. Which two ports are connected if the pilot pressure is 20 psig?

____

____

____________________ **5.** Is the switch point adjustable?

6. Why is the NC port of switching relay 2 switched to atmosphere?

____

____

____

7. What control causes switching relay 1 to switch? Under what condition(s)?

____

____

____

Activity 13-2. Electric/Pneumatic Switches

Locate all electric/pneumatic switches on the multizone air handling unit drawing and answer the questions.

_______________ **1.** How many electric/pneumatic switches are used in the multizone air handling unit?

2. List the electric/pneumatic switches.

____

____

____

3. If the electric/pneumatic solenoid is energized, which two ports are connected?

____

____

4. If the electric/pneumatic solenoid is de-energized, which two ports are connected?

____

____

5. If electric/pneumatic switch 2 is de-energized, what happens to the air at the common port? What does this cause at the humidifier valve?

____

____

____

6. If electric/pneumatic switch 1 is connected to the supply fan circuit, what happens at the outside air damper if the fan shuts down? Why would this be desirable?

____

____

____

Activity 13-3. Pneumatic/Electric Switches

Locate all pneumatic/electric switches on the multizone air handling unit drawing and answer the questions.

__________ **1.** How many pneumatic/electric switches are used in the multizone air handling unit?

2. List the pneumatic/electric switches.

____

____

____

3. Which electrical contacts are connected at 0 psig?

____

____

4. Which electrical contacts are connected at 20 psig?

____

____

__________ **5.** Do both pneumatic/electric switch 1 and pneumatic/electric switch 2 have the same setpoint?

6. If pneumatic/electric switch 1 controls the first stage cooling at one-third capacity and pneumatic/electric switch 2 controls the second stage cooling, estimate their setpoints.

____

____

____

7. Pneumatic/electric switch 3 is piped to the low-temperature limit control (thermostat). If the low-temperature limit thermostat bleeds the air off of the pneumatic/electric switch, what happens to the supply fan?

____

____

____

Activity 13-4. High Signal Selection Relay

Locate the high signal selection relay on the multizone air handling unit drawing and answer the questions.

1. What condition does the highest pressure value represent?

____

____

____________________ **2.** Should the zone thermostats be calibrated the same?

Activity 13-5. Auxiliary Device Troubleshooting

Scenario 1: A complaint that rooms are too hot has been received. The outside air temperature is 75°F. The highest zone pressure is 20 psig. Only stage 1 of cooling is running. A voltmeter placed across NO and C terminals of the second stage pneumatic/electric switch reads 110 VAC.

____________________ **1.** Is the switch open or closed?

2. What possible problem might this indicate?

____

____

____

Scenario 2: A complaint that rooms are too dry has been received. The outside air temperature is 30°F. The humidifier valve is closed. The humidity controller is calling for full humidity. There is 0 psig on the humidifier valve. An ohmmeter placed across the solenoid of electric/pneumatic switch 2 reads infinite resistance.

3. What is a possible problem with the solenoid?

____

____

____

Section 14.1 Pneumatic Control Diagrams

REVIEW QUESTIONS

Name ______________________________ **Date** ______________

Multiple Choice

______________ **1.** A(n) ___ is a drawing of a mechanical system that illustrates actual controls and piping between devices.
- A. parts list
- B. interface diagram
- C. control drawing
- D. sequence chart

______________ **2.** An electrical interface diagram is a drawing showing the interconnection between the pneumatic components and ___ equipment in a system.
- A. electronic
- B. electrical
- C. digital
- D. automated

______________ **3.** A(n) ___ is a drawing provided for a new system on installation.
- A. control drawing
- B. pneumatic control diagram
- C. as-built
- D. electrical interface diagram

______________ **4.** The ___ sequence must include the closing of the outside air dampers to the minimum position if the outside air conditions become unsuitable for cooling the building.
- A. damper/economizer
- B. zone temperature control
- C. humidification
- D. heating/cooling

______________ **5.** ___ are common problems that occur when using pneumatic control diagrams.
- A. Sequencing and design flaws
- B. Troubleshooting difficulties and sequencing flaws
- C. Pneumatic control diagram inaccessibility and overly simplified drawings
- D. Missing components and inaccurate pneumatic control drawings

______ **6.** A technician uses a(n) ___ to trace a control function for troubleshooting purposes.
A. sequence chart
B. interface diagram
C. control drawing
D. parts list

______ **7.** The entire ___ is required to obtain the correct replacement part needed in a pneumatic system.
A. pressure setpoint
B. part number
C. piping scheme
D. port number

______ **8.** The ___ sequence prevents building damage due to high humidity levels.
A. damper/economizer
B. zone temperature control
C. fire/smoke
D. humidification

______ **9.** Normally, ___ is included in a pneumatic control diagram.
A. only equipment interfaced directly to the pneumatic controls
B. only the electrical circuit connection
C. the auxiliary device
D. all equipment in an HVAC system

______ **10.** A ___ is a reference list that indicates part description acronyms and actual manufacturer part names and numbers.
A. parts list
B. electrical interface diagram
C. written sequence of operation
D. sequence chart

Completion

______ **1.** A(n) ___ diagram is a pictorial and written representation of pneumatic controls and related equipment.

______ **2.** Pneumatic control diagrams identify proper ___ of equipment, as well as potential trouble spots and design flaws.

______ **3.** A(n) ___ is a written description of the operation of a control system.

______ **4.** A(n) ___ is a chart that shows the numerical relationship between the different values in a pneumatic system.

______ **5.** A(n) ___ diagram is a drawing showing the interconnection between the pneumatic components and electrical equipment in a system.

Section 14.2 Pneumatic Control System Applications

REVIEW QUESTIONS

Name ______________________ **Date** ______________

Multiple Choice

______ **1.** The fan of a constant-volume air handling unit always runs at ___% of its rated capacity.
   A. 25
   B. 50
   C. 75
   D. 100

______ **2.** Most variable air volume air handling units maintain a constant static pressure of 1″ wc and a constant discharge air temperature of ___°F.
   A. 32
   B. 55
   C. 65
   D. 72

______ **3.** In pressure-dependent variable air volume systems, a thermostat controls the damper actuator directly without reference to the ___ of air flowing through the ductwork.
   A. volume
   B. pressure
   C. temperature
   D. humidity

______ **4.** A(n) ___ is an evaporative water cooler that uses natural evaporation to cool water.
   A. air chiller
   B. liquid chiller
   C. heat exchanger
   D. cooling tower

______ **5.** In ASHRAE Cycle ___, a fixed amount of outside air (normally 10% to 20%) is brought in during the heating cycle.
   A. I
   B. II
   C. III
   D. W

_______________ **6.** A(n) ___ is included to prevent the compressors in a single-zone air handling unit from operating if the supply fan is OFF.
A. electric switch
B. damper actuator
C. interlock
D. valve

_______________ **7.** ___ control is the most basic control method of any mechanical system.
A. Proportional/integral
B. Proportional
C. Single-zone
D. Multiple-zone

_______________ **8.** Multizone air handling units are identified by the number of ___ mounted on the discharge end of the air handler.
A. pressure switches
B. centrifugal pumps
C. air compressors
D. damper actuators

_______________ **9.** In ___ VAV systems, a reset flow controller is used to measure airflow through the VAV terminal box.
A. pressure-dependent
B. pressure-independent
C. temperature-dependent
D. temperature-independent

_______________ **10.** On a VAV air handling unit, the maximum setting on a reset flow controller prevents ___ through the VAV terminal box.
A. low air pressure
B. fresh air ventilation
C. discharge air
D. excessive noise

_______________ **11.** A ___ is a heat exchanger that removes heat from high-pressure refrigerant vapor.
A. condenser
B. cooling tower
C. liquid chiller
D. unit ventilator

_______________ **12.** The ___ opens the hot water valve and closes the outside air damper if the discharge air temperature is excessively low.
A. bypass air damper
B. low-limit thermostat
C. heat exchanger
D. condenser

______________ **13.** If multiple pumps are used in a boiler control application, the operation of the pumps may be regulated through a ___.
A. lead/lag switch
B. relay switch
C. VAV terminal box
D. unit ventilator control

Completion

______________ **1.** In pressure-independent variable air volume systems, a(n) ___ measures airflow through a variable air volume terminal box.

______________ **2.** A(n) ___ is a small air handling unit mounted on the outside wall of each room of a building.

______________ **3.** A(n) ___ air handling unit is an air handling unit that moves a constant volume of air.

______________ **4.** A(n) ___ is a system that uses a liquid (normally water) to cool building spaces.

______________ **5.** ___ normally provide comfort for only one space within a commercial building.

______________ **6.** A(n) ___ may be included in a cooling tower sump to keep the water temperature above freezing for winter operation.

______________ **7.** Variable air volume air handling units produce a constant ___°F air temperature.

Single-Zone Air Handling Unit

_______________ **1.** normally open return air damper

_______________ **2.** mechanical interlock

_______________ **3.** supply fan

_______________ **4.** zone thermostat

_______________ **5.** main air supply

_______________ **6.** normally closed exhaust air damper

_______________ **7.** electric/pneumatic switch

_______________ **8.** filter

_______________ **9.** supply air

_______________ **10.** exhaust air

_______________ **11.** damper actuator

_______________ **12.** return air

_______________ **13.** normally open hot water valve

_______________ **14.** outside air

_______________ **15.** minimum position relay

_______________ **16.** normally closed cold water valve

_______________ **17.** normally closed outside air damper

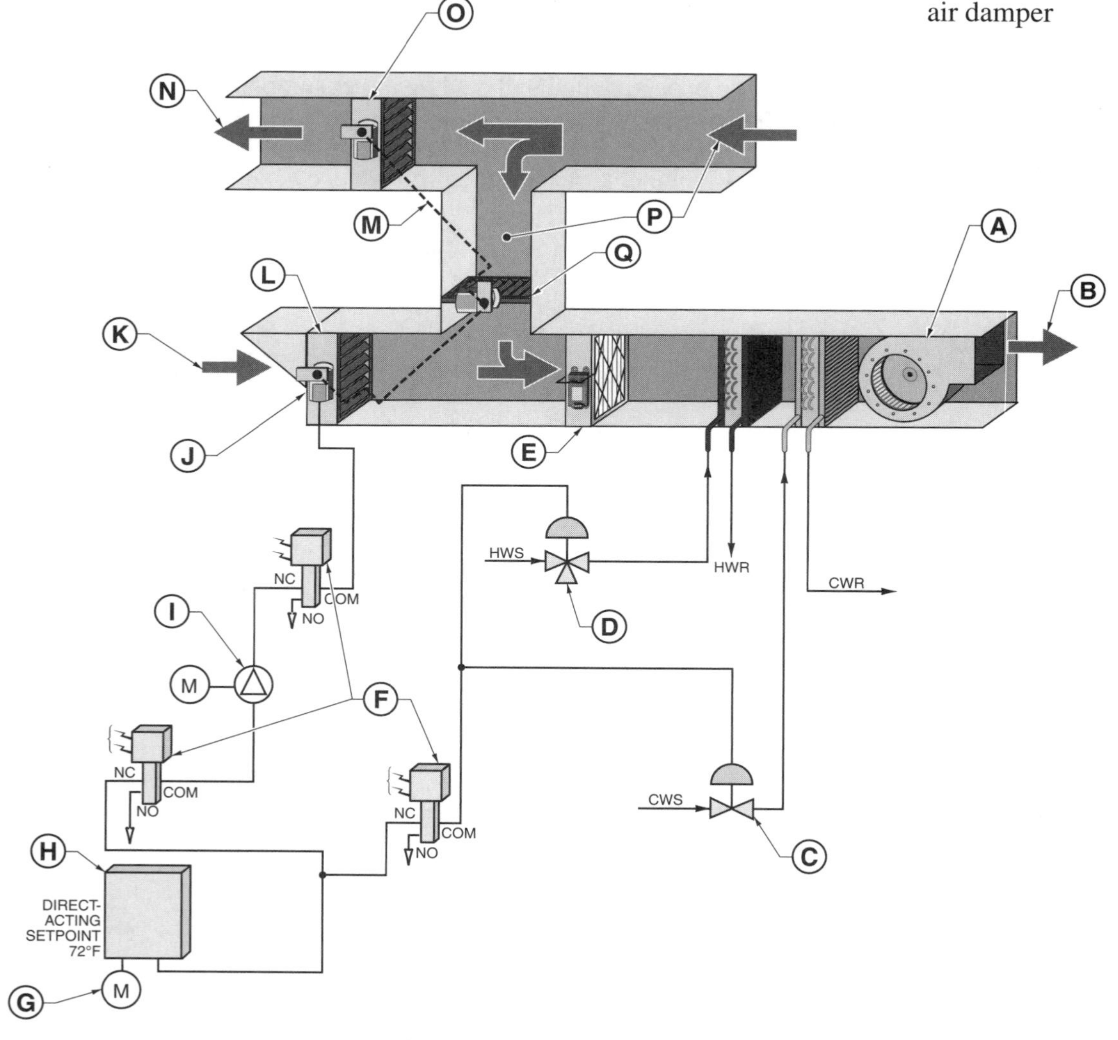

Variable-Air-Volume Air Handling Unit

______________ 1. switching relay
______________ 2. supply fan
______________ 3. damper actuators
______________ 4. duct static pressure transmitter
______________ 5. cooling coil
______________ 6. receiver controller (changeover control)
______________ 7. discharge air temperature transmitter
______________ 8. minimum position relay
______________ 9. normally closed outside air damper
______________ 10. three-way chilled water (mixing) valve
______________ 11. mixed air temperature transmitter
______________ 12. low-limit electric thermostat
______________ 13. filter
______________ 14. normally closed exhaust air damper
______________ 15. outside air transmitter
______________ 16. normally open return air damper

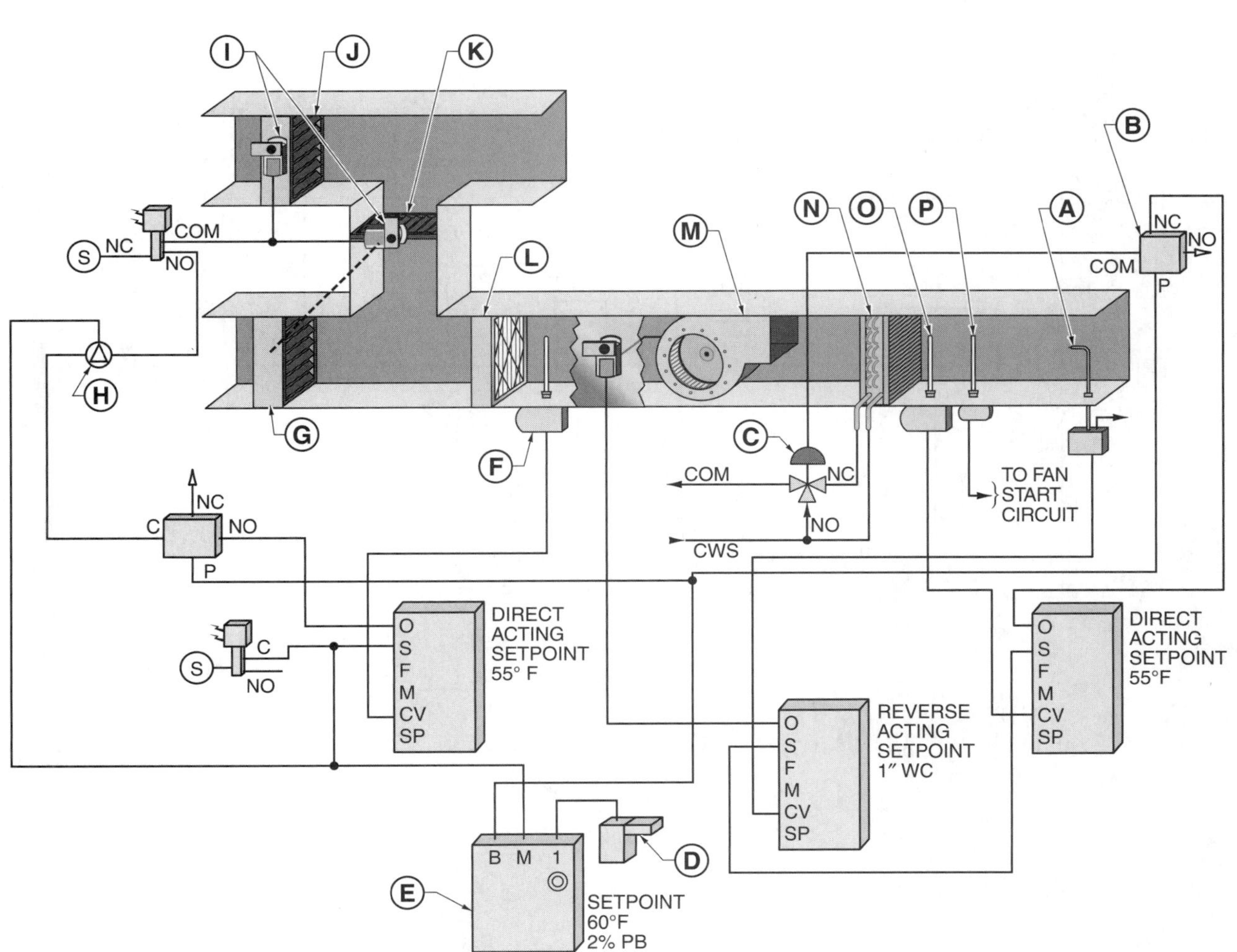

Boiler Control

______________	**1.** zone hot water valve
______________	**2.** boiler
______________	**3.** hot water return transmitter
______________	**4.** receiver controller (changeover control)
______________	**5.** heating/cooling coil
______________	**6.** three-way hot water valve
______________	**7.** restrictor
______________	**8.** circulating pump
______________	**9.** zone thermostat
______________	**10.** outside air transmitter
______________	**11.** pneumatic/electric switch
______________	**12.** receiver control (hot water control)
______________	**13.** panel receiver gauge
______________	**14.** hot water supply transmitter
______________	**15.** pump status differential pressure switch
______________	**16.** self-contained differential pressure actuator

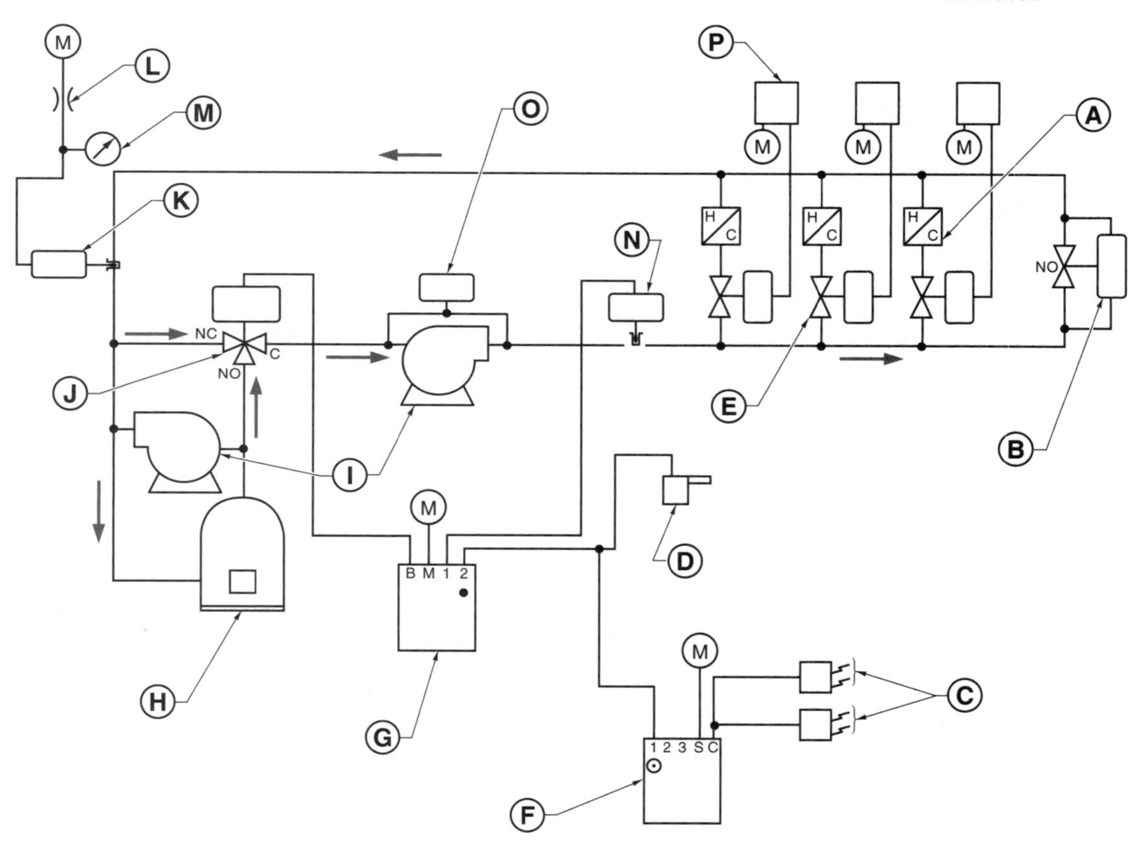

Pneumatic Control System Applications

ACTIVITIES

Name ______________________________ **Date** ______________________

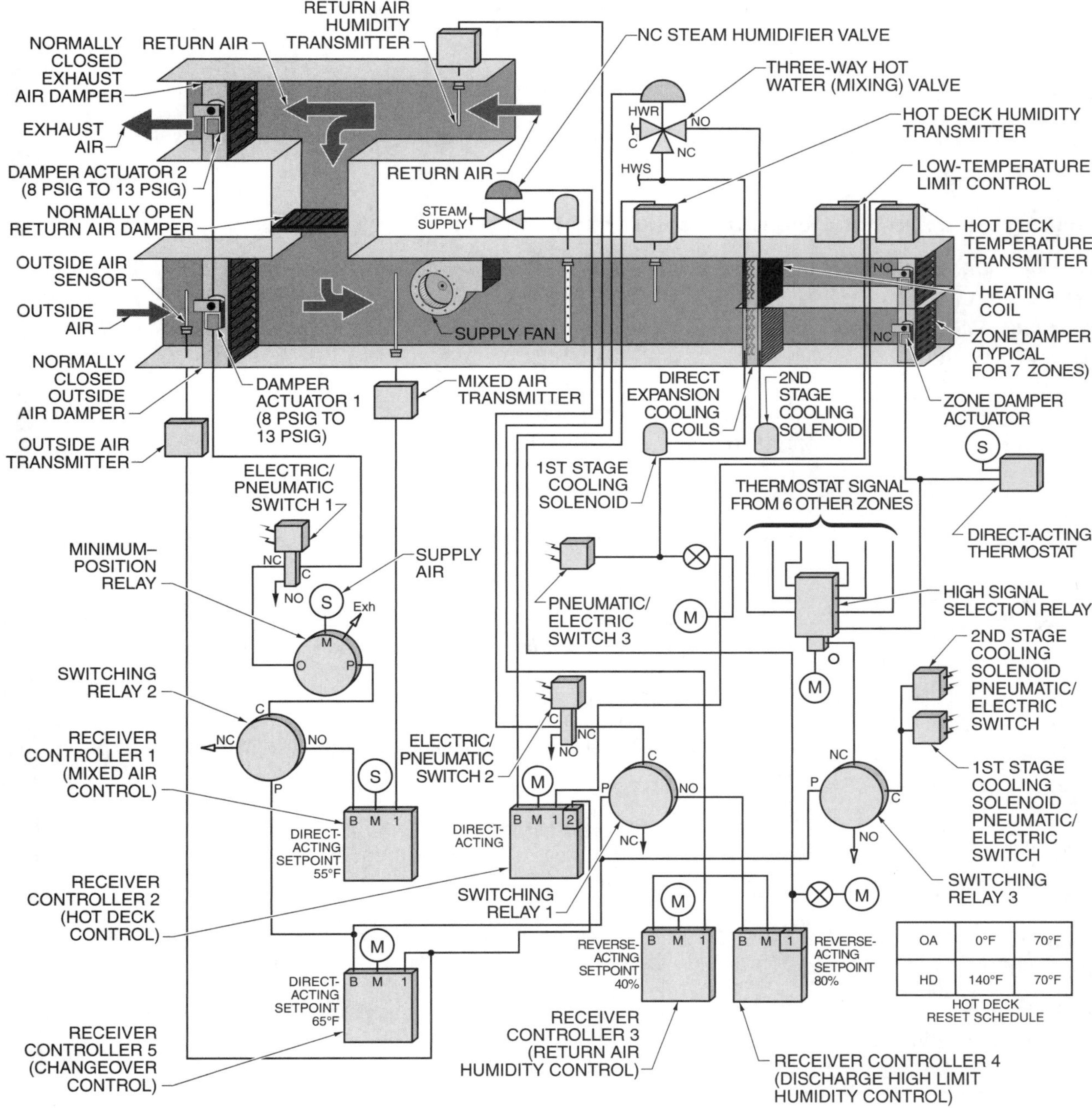

OA	0°F	70°F
HD	140°F	70°F

HOT DECK RESET SCHEDULE

Activity 14-1. Air Handling Unit Heating Control System

Use the multizone air handling unit drawing to answer the questions. Trace the heating controls of the system. The heating system controls include receiver controller 2 and the three-way hot water valve.

_______________ **1.** Are there any electric/pneumatic switches or switching relays between the output of receiver controller 2 and the three-way hot water valve?

_______________ **2.** If so, does their operation affect receiver controller 2 or the valve?

_______________ **3.** Which controller shares the outside air temperature sensor with the reset transmitter port of receiver controller 2?

_______________ **4.** If the outside air transmitter fails, does it affect both controllers?

_______________ **5.** Is the operation of the hot deck controller tied into the supply fan?

Activity 14-2. Air Handling Unit Cooling Control System

Use the multizone air handling unit drawing to answer the questions. Trace the cooling controls of the system. The cooling system controls include receiver controller 5, switching relay 3, first and second stage solenoid pneumatic/electric switches, high signal selection relay, first and second stage cooling solenoids, and zone thermostat.

_______________ **1.** If receiver controller 5 (changeover control) is operating properly, its branch pressure when the temperature is above 65°F is ___ psig.

_______________ **2.** At a temperature above 65°F, what two ports of switching relay 3 are connected?

_______________ **3.** Does the highest pressure from the signal selection relay then control the pneumatic/electric switches?

_______________ **4.** If receiver controller 5 (changeover control) is operating properly, its branch pressure is ___ psig when the temperature is below 65°F.

_______________ **5.** At a temperature below 65°F, what two ports of switching relay 3 are connected?

_______________ **6.** Does the highest pressure from the signal selection relay then control the pneumatic/electric switches?

_______________ **7.** The room with the highest demand for cooling controls the mechanical cooling equipment. If a room is calling for maximum cooling, would the damper for that zone be fully open to provide cool air to that zone?

_______________ **8.** Would the setpoints of the first and second stage pneumatic/electric switches be coordinated with the spring range of the damper actuator?

9. If this is true, and the zone damper spring range is 3 psig to 9 psig, give sample setpoints for the cooling solenoid pneumatic/electric switches.

____

____

Activity 14-3. Air Handling Unit Damper Control System

Use the multizone air handling unit drawing to answer the questions. Trace the damper controls of the system. The damper controls include receiver controller 1, receiver controller 5, switching relay 2, the minimum-position relay, electric/pneumatic switch 1, and damper actuator 1 and 2.

______________ **1.** The setpoint of receiver controller 1 (mixed air control) is ___°F.

______________ **2.** If receiver controller 5 (changeover control) is operating properly, its branch pressure is ___ psig when the temperature is above 65°F.

______________ **3.** At a temperature above 65°F, what two ports of switching relay 2 are connected?

4. At a temperature above 65°F, what happens to the air at the pilot port of the minimum position relay?

____

____

______________ **5.** At a temperature above 65°F, does the minimum position relay allow the outside air damper to fully close?

______________ **6.** For a given 8 psig to 13 psig spring range, what is a possible minimum position relay setpoint for 10% minimum outside air?

7. If electric/pneumatic switch 1 is tied into the fan operation, what happens to the outside air damper if the fan shuts down?

____

____

8. Why might this be desired?

____

____

______________ **9.** If receiver controller 5 (changeover control) is operating properly, its branch pressure is ___ psig at a temperature below 65°F.

______________ **10.** At a temperature below 65°F, what two ports of switching relay 3 are connected?

______________ **11.** Does receiver controller 1 then control the outside air, return air, and exhaust air dampers to maintain its setpoint?

Activity 14-4. Air Handling Unit Humidity Control System

Use the multizone air handling unit drawing to answer the questions. Trace the humidity controls of the system. The humidity controls include receiver controller 3, receiver controller 4, receiver controller 5, switching relay 1, electric/pneumatic switch 2, and the normally closed steam humidifier valve.

______________ **1.** If receiver controller 5 (changeover control) is operating properly, its branch pressure is ___ psig when the temperature is above 65°F.

____________ **2.** At a temperature above 65°F, what two ports of switching relay 1 are connected?

____________ **3.** Can the humidifier valve open when the outside air temperature is above 65°F?

____________ **4.** If electric/pneumatic switch 2 is connected to the operation of the supply fan, will the humidifier operate if the supply fan is off?

5. Why would this be desired?

6. Can mechanical cooling and the humidifier valve operate at the same time? Why or why not?

7. High receiver controller 4 (discharge high limit humidity control) has a setpoint of 80% rh. If the discharge humidity reaches a level above 80%, what does receiver controller 4 do to the steam humidifier valve?

8. Why would this be desired?

Activity 14-5. Air Handling Unit Low Temperature Limit Control System

Use the multizone air handling unit drawing to answer the questions. Trace the low temperature limit controls of the system. The low temperature limit controls include the low temperature limit control and pneumatic/electric switch 3.

____________ **1.** If it is desired to prevent a freezing condition after the heating coil, the low temperature limit would trip at ___°F.

2. When pneumatic/electric switch 3 stops the operation of the supply fan, what happens to the humidifier?

3. When pneumatic/electric switch 3 stops the operation of the supply fan, what happens to the outside air dampers?

____________ **4.** Is the operation of the hot deck controller tied into the low-temperature limit control?

Activity 14-6. Air Handling Unit Sequence of Operation

Use the single-zone air handling unit drawing to answer the questions. Circle the heating valve, cooling valve, and damper operation for your reference.

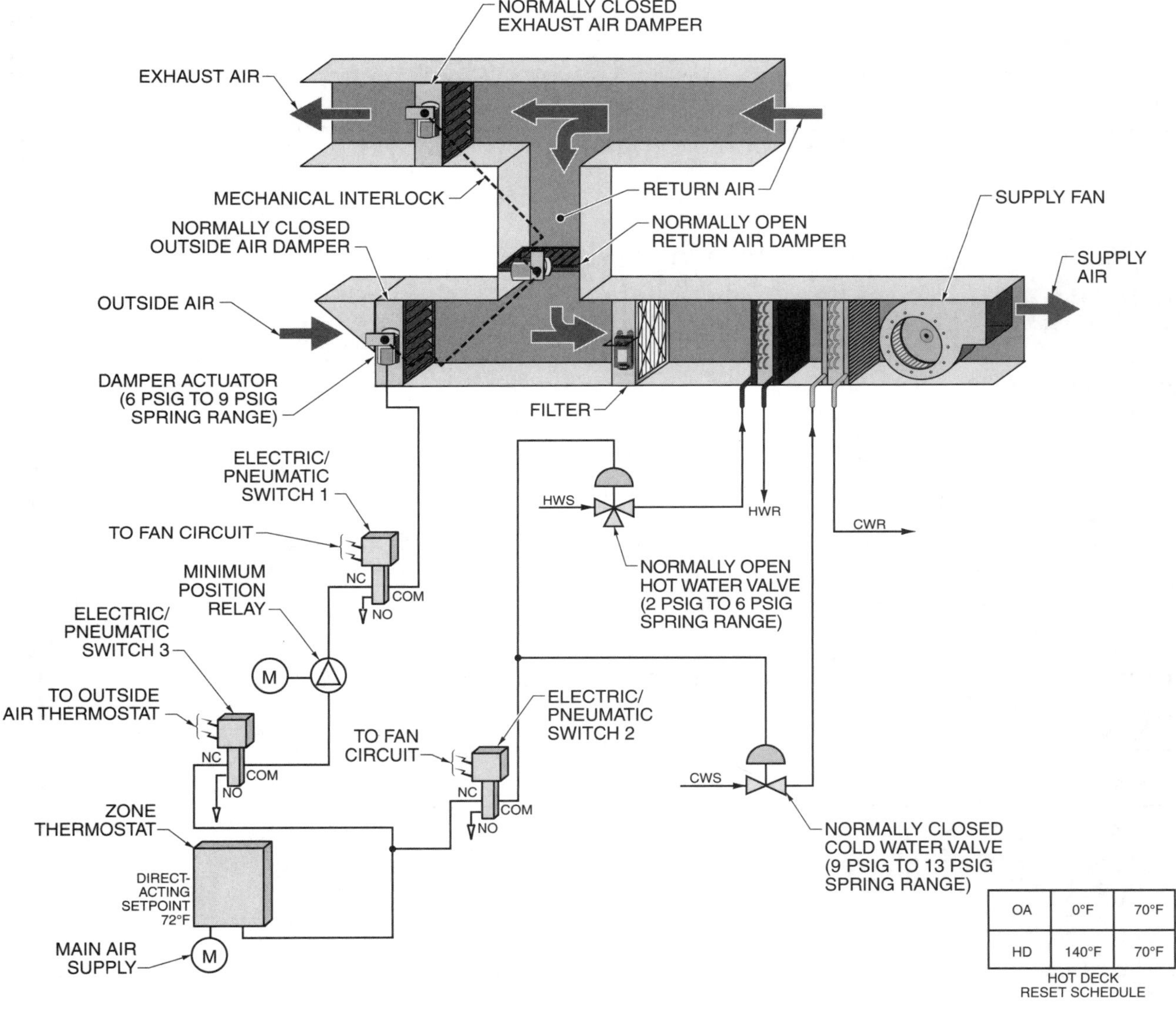

OA	0°F	70°F
HD	140°F	70°F

HOT DECK RESET SCHEDULE

1. Mark the operation of the heating valve, cooling valve, and damper actuator on the graph using their indicated spring ranges.

______________ 2. Is there any overlap between heating, cooling, and damper operations?

______________ 3. If spring range shift occurs for any of the devices, could there be simultaneous heating, cooling, or damper operation?

4. Write the heating sequence of operation for the single-zone air handling unit.

5. Write the cooling sequence of operation for the single-zone air handling unit.

6. Write the damper sequence of operation for the single-zone air handling unit.

Section 15.1 Electricity and Controls

REVIEW QUESTIONS

Name ______________________________ **Date** ______________________

Multiple Choice

______________ **1.** Electricity is the energy released by the flow of ___ in a conductor (wire).

A. neutrons
B. protons
C. electrons
D. current

______________ **2.** ___ is the amount of electrons flowing through a conductor.

A. Resistance
B. Voltage
C. Current
D. Electricity

______________ **3.** A ___ is a switch that isolates electrical circuits from the voltage source to allow safe access for maintenance or repair.

A. relay
B. DPDT switch
C. pressure switch
D. disconnect

______________ **4.** A hygroscopic element is a device that changes its characteristics as ___ changes.

A. humidity
B. temperature
C. current
D. voltage

______________ **5.** A(n) ___ control system is a control system that uses electricity (24 VAC or higher) to operate devices in the system.

A. mechanical
B. electrical
C. high-limit
D. packaged unit

______________ **6.** An insulator is a material that has a high ___ and resists the flow of electrons.

A. resistance
B. voltage
C. current
D. electrical charge

______________ **7.** ___ is the positive (+) or negative (–) state of an object.
A. Voltage
B. Resistance
C. Current
D. Polarity

______________ **8.** Control systems at ___ VAC are the most common line-voltage control systems.
A. 95
B. 105
C. 120
D. 160

______________ **9.** A ___ is an electrically operated switch that includes motor overload protection.
A. contactor
B. motor starter
C. circuit breaker
D. relay

Completion

______________ **1.** ___ is the international unit of frequency equal to one cycle per second.

______________ **2.** Electrical control systems at ___ are the most common line-voltage control systems.

______________ **3.** A(n) ___ is a device that shuts OFF the power supply when current flow is excessive.

______________ **4.** ___ is the increasing of a controlled variable above the controller setpoint.

______________ **5.** A(n) ___ can be used to ensure proper air and water flow by measuring the differential pressure across a fan or pump.

______________ **6.** A(n) ___ is a mechanical procedure that consists of reversing refrigerant flow in a system to melt ice that builds up on the evaporator coil.

______________ **7.** ___ is current that reverses its direction of flow at regular intervals.

______________ **8.** A(n) ___ transformer is a transformer in which the secondary coil has fewer turns of wire than the primary coil.

______________ **9.** A(n) ___ is a device that prevents overcurrent in an electrical circuit.

Contactors and Motor Starters

_______________ **1.** 3ϕ power supply lines

_______________ **2.** motor power wires

_______________ **3.** normally closed auxiliary contact terminals

_______________ **4.** coil terminals

_______________ **5.** overload contact

_______________ **6.** power supply terminals

_______________ **7.** coil cover

_______________ **8.** secondary terminals

_______________ **9.** auxiliary contact

_______________ **10.** heaters

_______________ **11.** coil

_______________ **12.** normally open auxiliary contact terminals

_______________ **13.** auxiliary contact manual activation button

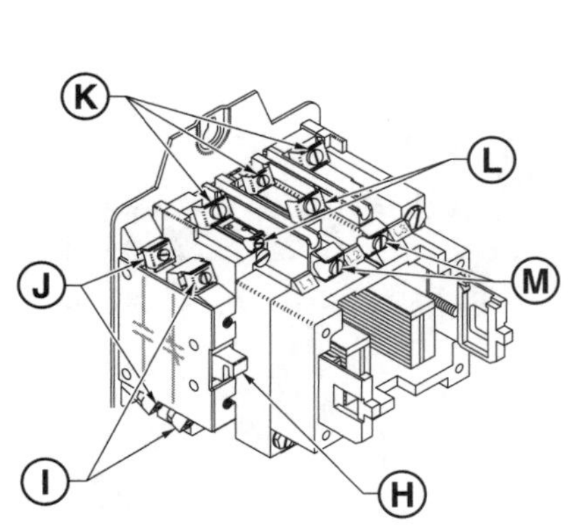

CONTACTOR

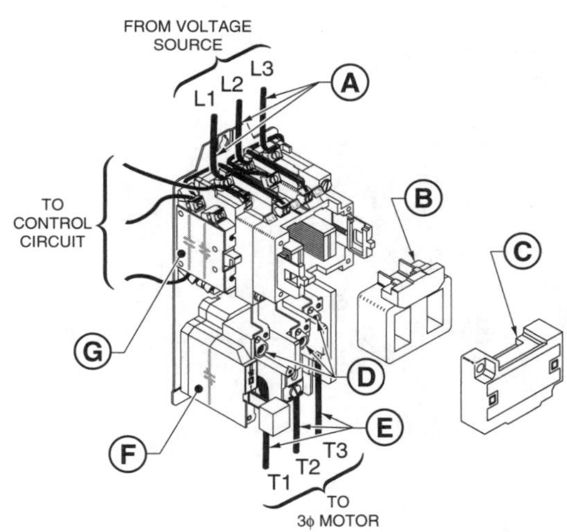

MAGNETIC MOTOR STARTER

Series Circuit Connections

_____________ 1. relay contact coil

_____________ 2. low pressure switch

_____________ 3. current flow

_____________ 4. one path for current flow

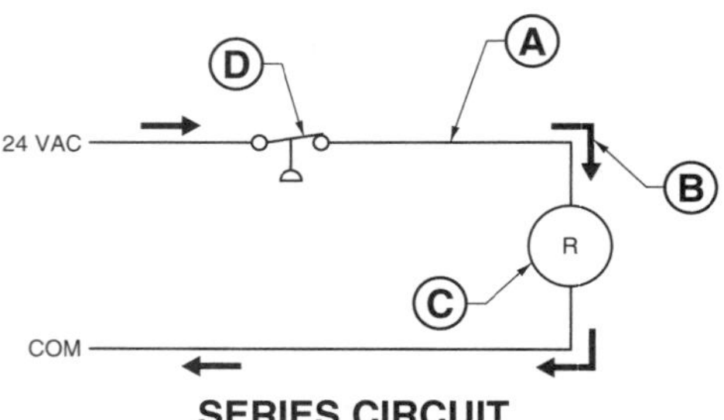

SERIES CIRCUIT

Pressure Switches

_____________ 1. metal bellows

_____________ 2. neoprene diaphragm

_____________ 3. atmospheric pressure

_____________ 4. output to mechanical linkage and switch

_____________ 5. system pressure

_____________ 6. output force to linkage

_____________ 7. system pressure connection

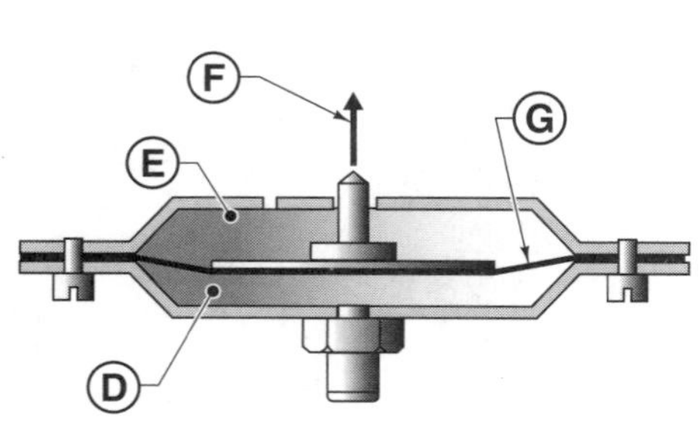

DIAPHRAGM ELEMENT

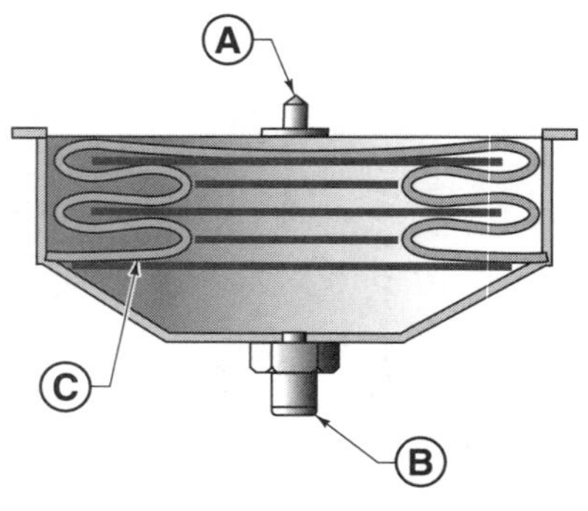

BELLOWS ELEMENT

Section 15.2 Electrical Control System Applications

REVIEW QUESTIONS

Name ______________________________ **Date** ____________________

Multiple Choice

________________ **1.** ___ is measured across an electrical control device by placing the DMM leads in parallel across the device being tested.

A. Pressure
B. Resistance
C. Current
D. Voltage

________________ **2.** A(n) ___ is current that leaves the normal current-carrying path by going around the load and back to the power source or to ground.

A. dead short
B. short circuit
C. partial short
D. open circuit

________________ **3.** In a cooling control application, when the thermostat is in the OFF position, the ___.

A. compressor is ON, and the condenser fan and indoor fan are OFF
B. compressor is OFF, and the condenser fan and indoor fan are ON
C. compressor, condenser fan, and indoor fan are all ON
D. compressor, condenser fan, and indoor fan are all OFF

________________ **4.** In a refrigeration control application, when power is applied to the HVAC unit, the ___ is energized.

A. evaporator fan
B. compressor
C. compressor crankcase heater
D. liquid line solenoid valve

________________ **5.** A digital multimeter provides an audible signal to indicate ___.

A. a complete electrical path between the points tested
B. high resistance
C. incomplete electrical continuity between the points tested
D. a small number of electrons flowing through the circuit

____________________ **6.** In a(n) ___ application, a room thermostat energizes a gas valve, which allows gas to flow to a burner.
A. sequence timing
B. electrical
C. cooling control
D. heating control

Completion

____________________ **1.** ___ is the presence of a complete path for current flow.

____________________ **2.** In HVAC, all applicable safety codes and regulations must be followed per the ___ and/or authority having jurisdiction.

____________________ **3.** A(n) ___ is a device designed to measure current in a circuit by measuring the strength of the magnetic field around a single conductor.

____________________ **4.** A(n) ___ is a short circuit of only a section or several sections of a machine.

____________________ **5.** A(n) ___ is a test tool used to measure two or more electrical values.

Name ____ **Date** ________________________

Activity 15-1. Relay Troubleshooting

It is summer. A too-hot complaint is received. Upon arrival in the room, it is determined that the indoor fan is running but the compressor and outdoor fan motor are not. The thermostat fan switch is in the auto position. A digital multimeter (DMM) set to measure voltage is used to check the voltage in the circuit.

________________ **1.** With the indoor fan operating normally, the voltage across IFR1 is ___ VAC.

________________ **2.** With the compressor and outdoor fan motor not operating, the voltage across CC1 is ___ VAC.

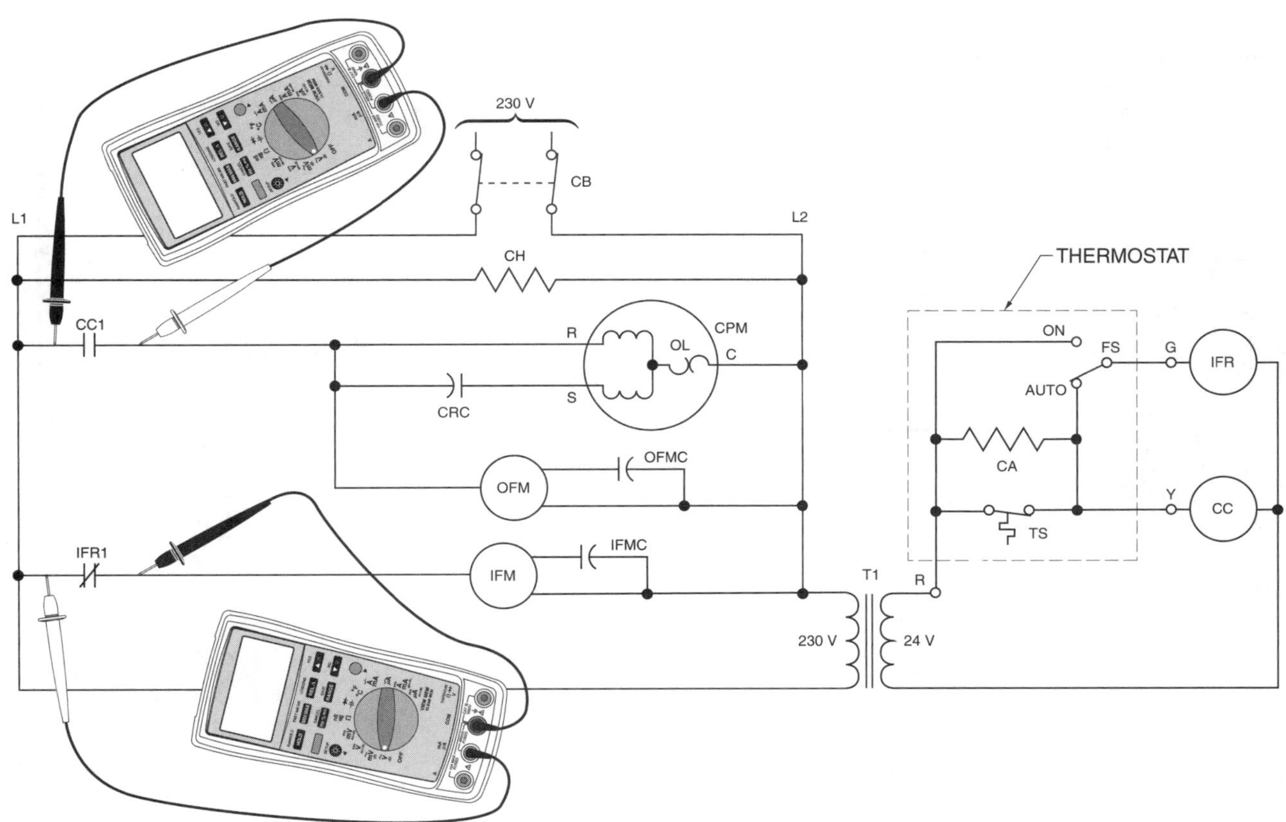

Activity 15-2. Transformer Secondary Troubleshooting

It is summer. A too-hot complaint is received. Upon arrival in the room, it is determined that the indoor fan, compressor, and outdoor fan motor are not running. The thermostat fan switch is in the auto position. A DMM set to measure voltage is used to check the voltage across the circuit breaker. The voltage level is okay. The control transformer primary is checked and the voltage level is 230 VAC. The voltage across the control transformer secondary coil is also checked.

_______________ **1.** The voltage across the control transformer secondary coil when it is open is ___ VAC.

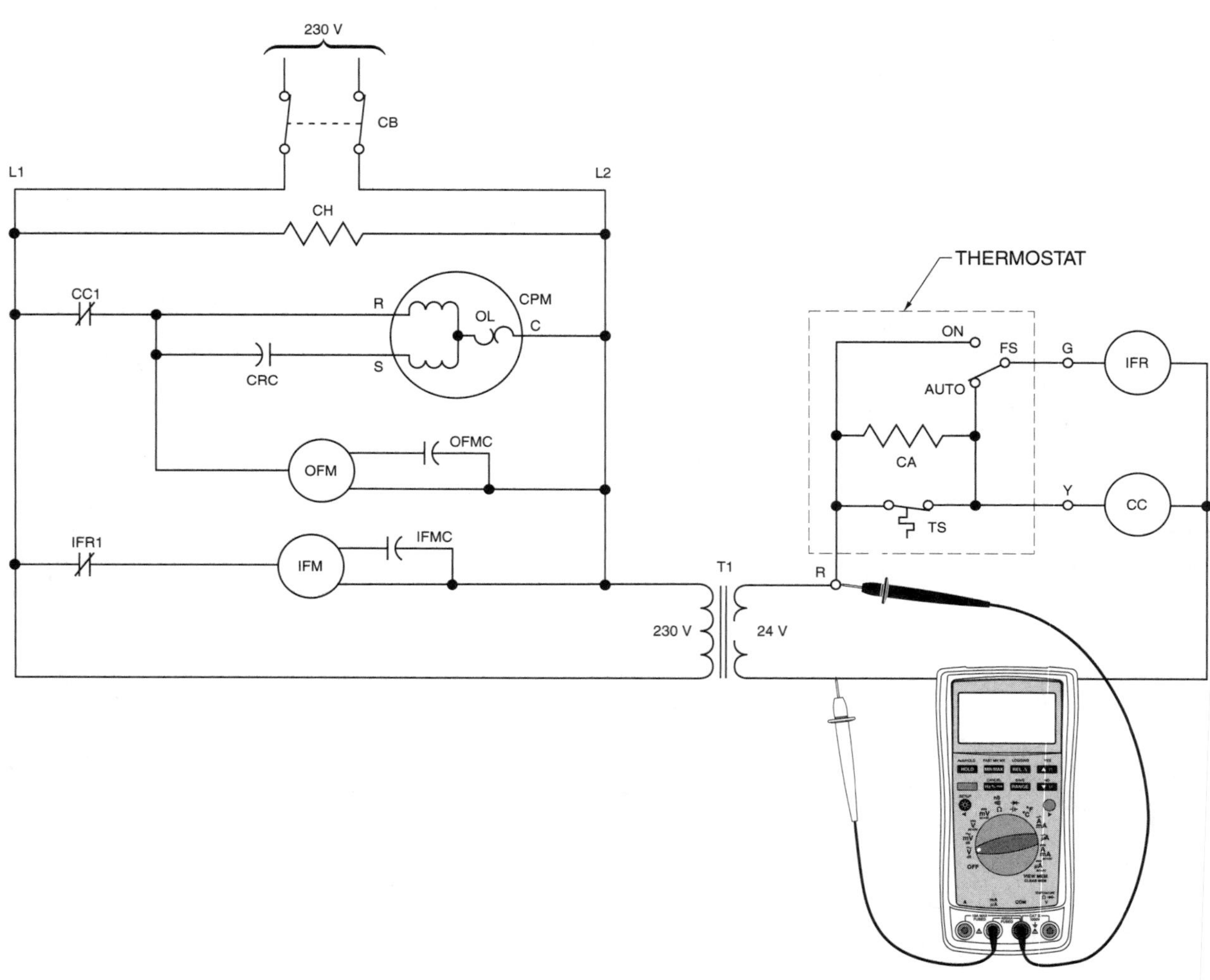

Activity 15-3. Thermostat Troubleshooting

It is summer. A too-cold complaint is received. Upon arrival in the room, it is determined that the indoor fan, compressor, and outdoor fan motor are running. The thermostat is set at 74°F, but it is 68°F in the room. The thermostat fan switch is in the auto position. Regardless of the temperature at which the thermostat is set, the unit never shuts off. A bad thermostat is suspected. After locking and tagging out the power supply, a DMM set to measure resistance is used to check the thermostat.

_______________ **1.** The resistance value of the thermostat if its contacts are closed is ___ Ω.

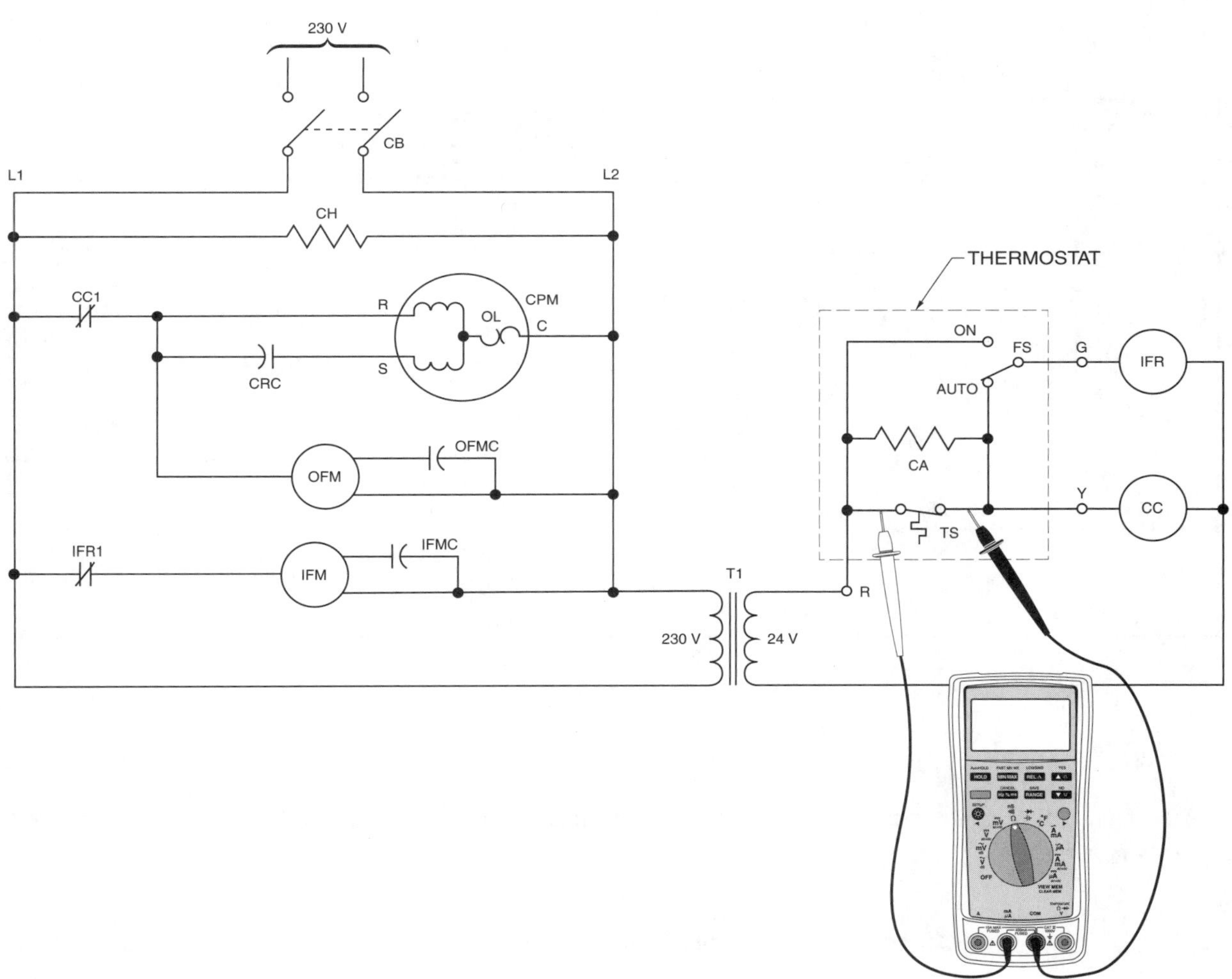

Activity 15-4. Circuit Current Troubleshooting

It is summer. A too-hot complaint is received. The only diagnostic tool available is a digital clamp-on ammeter. Upon arrival, it is determined that the indoor fan is running but the compressor and outdoor fan motor are not. The full-load current of the indoor fan is 3 A, the compressor is 20 A, and the outdoor fan is 2 A. The thermostat fan switch is in the auto position. The jaws of the ammeter are clamped around the conductors in the circuit.

____________________ **1.** The current reading of the clamp-on ammeter based on the these conditions is ___ A.

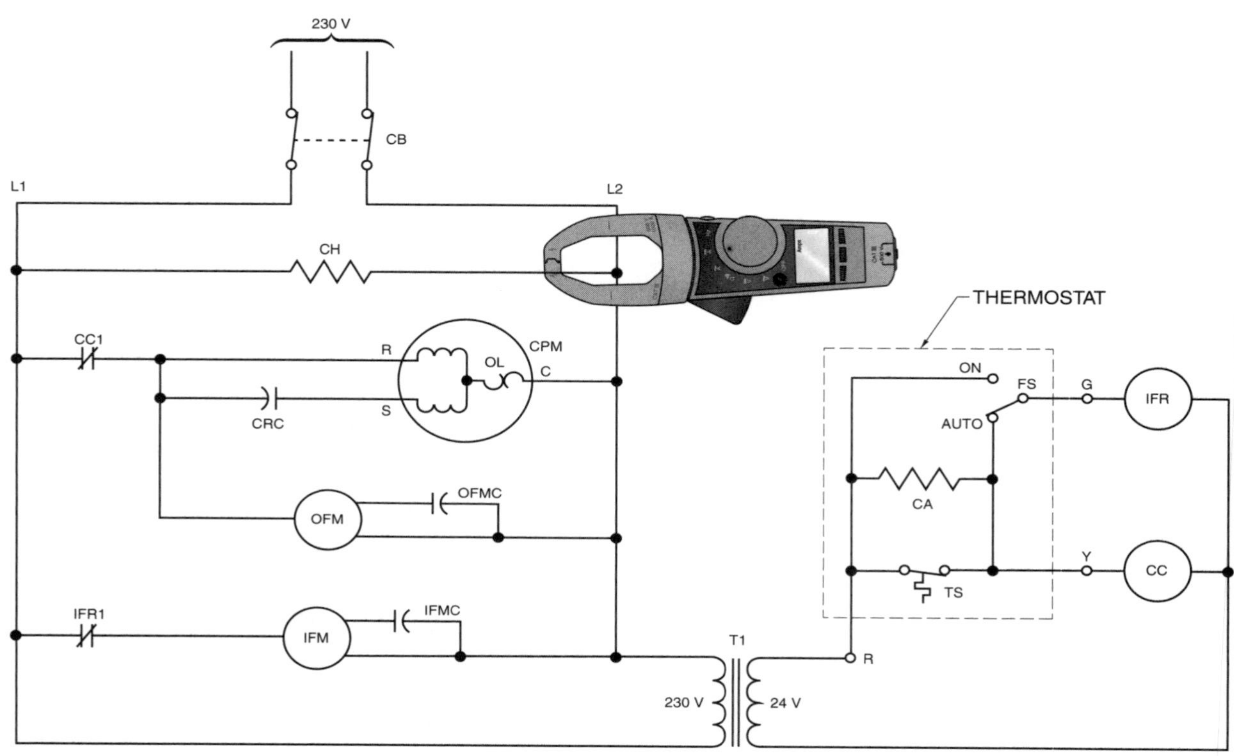

It is discovered that the thermostat was set to 80°F. After adjusting the thermostat to 74°F, the compressor and outdoor fan energize.

____________________ **2.** The current reading of the clamp-on ammeter based on the new conditions is ___ A.

Section 16.1 Electronic Control Systems

REVIEW QUESTIONS

Name ______________________________ Date ____________________

Multiple Choice

______________ **1.** ___ allow electrons to flow to plates that are located inside a glass tube.
A. Vacuum tubes
B. Transistors
C. Rectifiers
D. Light-emitting diodes

______________ **2.** A rectifier is a device that changes AC voltage into ___ voltage.
A. three-phase
B. single-phase
C. DC
D. AC

______________ **3.** A bridge rectifier is a circuit containing ___ diodes that permit both halves of the AC sine wave to pass.
A. one
B. two
C. three
D. four

______________ **4.** A ___ rectifier is a circuit containing one diode that allows only half of the input AC sine wave to pass.
A. bridge
B. half-wave
C. full-wave
D. field-effect

______________ **5.** A triac is a solid-state device used to switch ___.
A. direct current
B. alternating current
C. devices with small resistance
D. devices with high resistance

______________ **6.** An inexpensive ___ worn by technicians and connected to ground is used to protect against static electrical discharge to electronic components.
A. glove
B. wrist strap
C. trace
D. semiconductor device

______________ **7.** The most common transistors used today are ___ transistors.
A. field-effect
B. junction field-effect
C. metal-oxide semiconductor field-effect
D. bridge

______________ **8.** An electronic control system is a control system in which the power supply is ___ VDC or less.
A. 8
B. 12
C. 16
D. 24

Completion

______________ **1.** ___-type material is material created by doping a region of a crystal with atoms from an element that has more electrons in its outer shell than the crystal.

______________ **2.** A(n) ___ is a material in which electrical conductivity is between that of a conductor (high conductivity) and that of an insulator (low conductivity).

______________ **3.** A(n) ___ is a semiconductor device that allows current to flow in one direction only.

______________ **4.** A(n) ___ is a circuit containing two diodes and a center-tapped transformer that permits both halves of the input AC sine wave to pass.

______________ **5.** A(n) ___ diode is commonly used as a voltage shunt or electronic safety valve.

______________ **6.** A(n) ___ is a three-terminal semiconductor device that controls current flow according to the amount of voltage applied to the base.

______________ **7.** A(n) ___ transistor is a transistor that consists of a phototransistor and a standard NPN transistor in a single package.

______________ **8.** A(n) ___ acts as a solid-state relay that can switch devices such as heaters, compressors, motors, and relays ON and OFF.

______________ **9.** A(n) ___ is an electronic device in which all components (transistors, diodes, and resistors) are contained in a single package or chip.

______________ **10.** A(n) ___ is an electronic device that changes resistance or switches ON when exposed to light.

______________ **11.** ___ is the addition of material to a base element to alter the crystal structure of the element.

______________ **12.** A silicon-controlled rectifier (SCR) is a thyristor that is capable of switching ___ current.

______________ **13.** ___-type material is material created by doping a region of a crystal with atoms from an element that has fewer electrons in its outer shell than the crystal.

______________ **14.** A(n) ___ is a diode designed to produce light when forward biased.

Photodiodes

______________ **1.** cathode lead

______________ **2.** metal can

______________ **3.** photodiode chip

______________ **4.** window

______________ **5.** anode lead

Light-Emitting Diodes

______________ **1.** anode lead

______________ **2.** epoxy housing

______________ **3.** flat mark

______________ **4.** cathode lead

______________ **5.** LED chip

______________ **6.** reflector

Bipolar Junction Transistor Construction

____________________ 1. emitter

____________________ 2. N-type material

____________________ 3. base

____________________ 4. P-type material

____________________ 5. collector

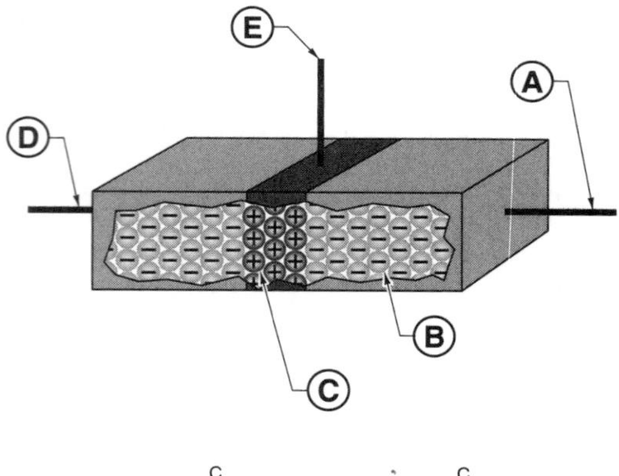

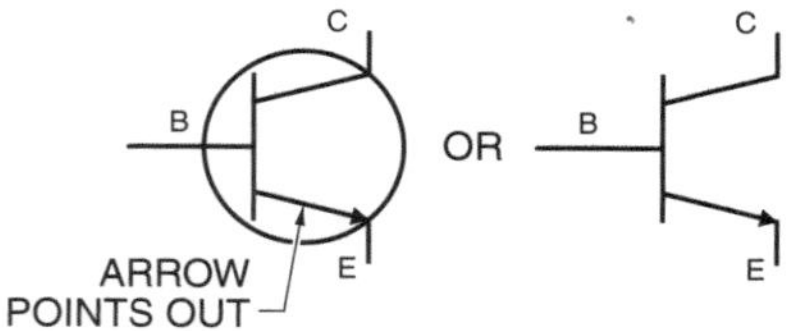

NPN

Metal-Oxide Semiconductor Field-Effect Transistor Construction

____________________ 1. drain

____________________ 2. P-type material

____________________ 3. gate

____________________ 4. metal-oxide material

____________________ 5. source

____________________ 6. insulator

____________________ 7. N-type material

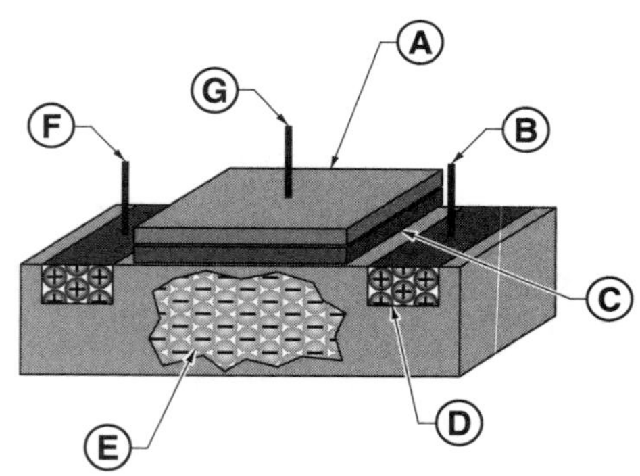

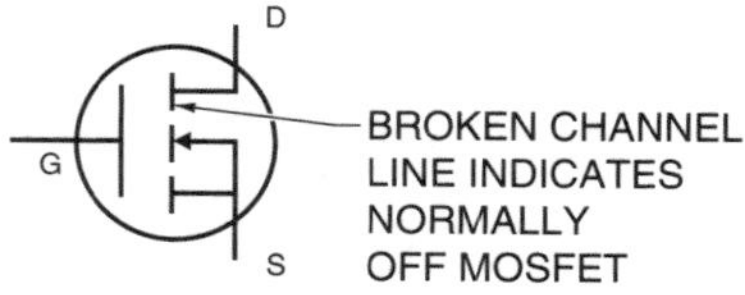

N-CHANNEL

Section 16.2 Electronic Control System Applications

REVIEW QUESTIONS

Name ______________________________ **Date** ______________________

Multiple Choice

______________ **1.** Electronic control systems include the sensors, controllers, switching components, and output devices such as dampers and valve actuators to provide ___ control in the building.

A. multizone unit
B. temperature, pressure, and humidity
C. boiler and chiller
D. grounding and electrical output

______________ **2.** A piezoelectric crystal is used by a(n) ___ sensor.

A. temperature
B. humidity
C. electronic
D. pressure

______________ **3.** In multizone unit control, each zone damper is controlled by a(n) ___.

A. zone thermostat
B. integrated circuit
C. air handling unit
D. electronic controller

______________ **4.** An electronic controller changes the hot water setpoint from 100°F to ___°F as the outside air temperature changes from 65°F to –10°F.

A. 25
B. 87
C. 190
D. 213

______________ **5.** An electronic controller changes the hot water setpoint from 100°F to ___°F as the outside air temperature changes from 65°F to –10°F.

A. 55
B. 60
C. 72
D. 85

Completion

_______________ **1.** ___ sensors use semiconductor materials that change resistance characteristics as the temperature around the sensor changes.

_______________ **2.** ___ sensors use a piezoelectric crystal, which changes output voltage as pressure is exerted on the crystal.

_______________ **3.** The electronic controller opens and closes the ___ electronic actuator to maintain the correct chilled water supply or return temperature.

_______________ **4.** Electronic thermostats use ___ to energize and de-energize heating, cooling, and fan functions.

ACTIVITIES

Name ______________________________ Date ____________________

Activity 16-1. Operator Interfaces

A too-hot complaint is received from one of the rooms in a building. The LED display on a building automation system operator interface is used to determine which temperature sensor is in alarm and its value. Use the operator interface information to answer the questions.

______________ **1.** Is the X LED lit?

2. What does this indicate?

____

______________ **3.** What is the value of the left LEDs?

______________ **4.** The left LEDs indicate input number ___.

______________ **5.** The right LEDs indicate the number ___.

______________ **6.** If this is a temperature sensor, its value is ___.

______________ **7.** Is the AL LED lit?

8. What does this mean?

____

____

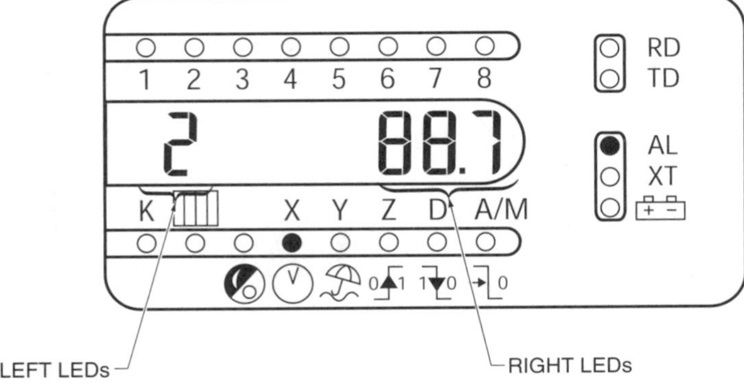

X = Analog Input Mode
AL = Point Is in Alarm
LEFT LEDs = Item Number
RIGHT LEDs = Value

Use the operator interface information to answer the questions.

____________________ **9.** Is the X LED lit?

10. What does this indicate?

____

____

____________________ **11.** What is the value of the left LEDs?

____________________ **12.** The left LEDs indicate input number ___.

____________________ **13.** The right LEDs indicate the number ___.

____________________ **14.** If this is a temperature sensor, its value is ___.

____________________ **15.** Is the AL LED lit?

16. What does this mean?

____

____

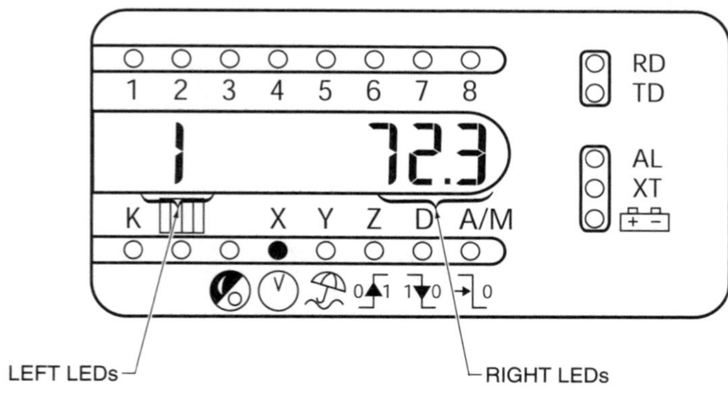

X = Analog Input Mode
AL = Point Is in Alarm
LEFT LEDs = Item Number
RIGHT LEDs = Value

Activity 16-2. Electronic Circuit Troubleshooting

A too-cold complaint is received for one of the rooms in a building. Upon arrival in the mechanical room, it is determined that the air handling unit fan is off. Use the print to answer the questions.

____________________ **1.** Is the device labeled A a thyristor?

____________________ **2.** What type of device is it?

____________________ **3.** Is the device in an AC or DC circuit?

____________________ **4.** The voltage level of the circuit is ___.

____________________ **5.** What device interfaces the low-voltage circuit to the line-voltage circuit?

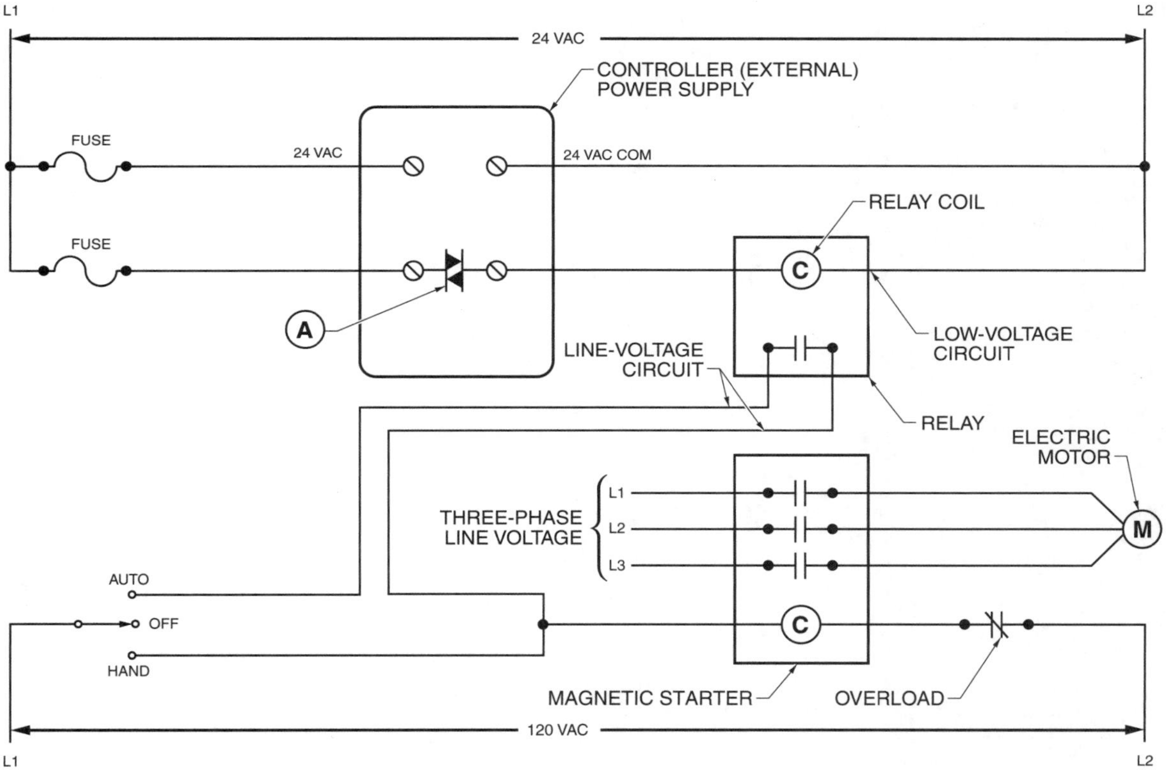

Section 17.1 Building Automation Systems

REVIEW QUESTIONS

Name ______________________________ Date ____________________

Multiple Choice

______________ **1.** A ___ system is a control system in which the decision-making equipment is located in one place and the system enables/disables local (primary) controllers.

A. central supervisory control
B. building automation
C. central-direct digital control
D. distributed direct digital control

______________ **2.** Some manufacturers provide ___ that allow central supervisory control systems to communicate with modern building automation system networks.

A. unitary controllers
B. universal input-output controllers
C. gateway interface modules
D. network communication modules

______________ **3.** Central-direct digital control systems are typically used for ___ instead of closed loop control.

A. load control
B. module control
C. supervisory control
D. sensor control

______________ **4.** A(n) ___ is a notification of improper temperature or other condition existing in a building.

A. setpoint
B. network communication point
C. communication bus
D. alarm

______________ **5.** A control function of a(n) ___ control system includes clock functions, humidity control, and pressure control.

A. central supervisory
B. application-specific
C. central-direct digital
D. distributed direct digital

______________ **6.** A(n) ___ control is control in which feedback occurs between the controller sensor and controlled device.
A. closed loop
B. open loop
C. distributed
D. direct digital

______________ **7.** ___ may have limited or no decision-making ability, which is the reason they may be referred to as dumb panels or dumb inputs/outputs (I/Os).
A. Stand-alone controls
B. Temperature sensors
C. Field interface devices
D. Unitary controller

______________ **8.** ___ control systems do not control a load directly, but enable or disable exisiting controllers.
A. Central-direct digital
B. Central supervisory
C. Application-specific
D. Distributed-direct digital

Completion

______________ **1.** A(n) ___ system is a system that uses microprocessors (computer chips) to control the energy-using devices in a building.

______________ **2.** A(n) ___ is an electronic device that follows commands sent to the device from the CPU of a central-direct digital control system.

______________ **3.** ___ is the recording of information such as temperature and equipment ON/OFF status at regular intervals.

______________ **4.** A(n) ___ control system is a control system that has multiple CPUs at the controller level.

______________ **5.** ___ is data needed by all controllers in a network.

______________ **6.** ___ is the generating of forms to notify maintenance personnel of routine maintenance procedures.

______________ **7.** In a central supervisory control system, all of the control decisions are made by a(n) ___ centrally located in the building or complex.

______________ **8.** ___ control systems are modular and easily expandable.

______________ **9.** A system-wide failure occurs if the ___ of a central supervisory control system fails.

______________ **10.** ___ is data required by all controllers in a network and includes outside air temperature and electrical demand.

______________ **11.** ___ control is less risky than other types of control because a controller failure is local and has minimal effect on the entire system.

______________ **12.** A(n) ___ control system is a control system in which all decisions are made in one location and which provides closed loop control.

REVIEW QUESTIONS

Name ________________________________ Date ____________________

Multiple Choice

______________ **1.** A(n) ___ is a controller designed to control only one type of HVAC system.

A. application-specific controller (ASC)
B. unitary controller
C. universal input-output controller
D. network communication controller

______________ **2.** A(n) ___ controller is a controller designed to control most HVAC equipment.

A. variable air volume box
B. air handling unit
C. network thermostat
D. universal input-output

______________ **3.** A ___ controller is a controller designed for basic zone control using a standard wall-mount temperature sensor.

A. variable air volume box
B. unitary
C. universal input-output
D. field interface

______________ **4.** A(n) ___ is a controller that coordinates communication from controller to controller on a network and provides a location for an operator interface.

A. application-specific controller
B. unitary controller
C. network communication module
D. variable air volume box controller

______________ **5.** A web-based interface module uses a ___ to access the building automation system from anywhere on the network.

A. secure log-in
B. private virtual network
C. standard web browser
D. human-computer interface

_______________ **6.** A common application of a network thermostat controller is for the control of a ___.
- A. heat pump
- B. hot water loop application
- C. VAV terminal box
- D. central-station air handling unit

_______________ **7.** An air volume increase is produced by controlling the speed of the fan motor using a(n) ___ or by opening dampers on the fan to admit a greater amount of air.
- A. universal input-output controller (UIOC)
- B. network communication module (NCM)
- C. air handling unit (AHU)
- D. variable-frequency drive (VFD)

_______________ **8.** Most universal input-output controllers use a(n) ___ VAC power supply.
- A. 8
- B. 16
- C. 24
- D. 36

Completion

_______________ **1.** ___ is the physical parts that make up a device.

_______________ **2.** Factory installation of application-specific controllers by a manufacturer is referred to as ___.

_______________ **3.** A single ___ controller may be used to control lighting and hot water applications as long as capacity of the controller is not exceeded.

_______________ **4.** ___ is the program that enables a controller to function.

_______________ **5.** A(n) ___ controller is a controller that modulates the damper inside a variable air volume (VAV) terminal box to maintain a specific building space temperature.

_______________ **6.** The term "___" is often used to describe network communication modules because they require software that is only provided by their manufacturer to access them.

_______________ **7.** In the water source heat pump application, the heat pump reversing valve is cycled by the ___.

_______________ **8.** One advantage of a(n) ___ module is that the software is often embedded in the device and is accessed through a standard web browser.

_______________ **9.** Unitary controllers are normally compact to allow mounting at the ___.

_______________ **10.** ___ controllers are custom programmed in the field.

_______________ **11.** Pressure-___ VAV terminal boxes are basic VAV terminal boxes that control air volume only.

_______________ **12.** A(n) ___ is a controller that contains inputs and outputs required to operate large central-station air handling units.

Unitary Controller Rooftop Unit

______________________ **1.** room temperature sensor

______________________ **2.** exhaust air

______________________ **3.** unitary controller

______________________ **4.** heating and cooling stages

______________________ **5.** return air

______________________ **6.** outside air temperature sensor

______________________ **7.** supply fan

______________________ **8.** airflow switch

______________________ **9.** economizer damper actuator

______________________ **10.** supply air

______________________ **11.** outside air

______________________ **12.** supply air temperature sensor

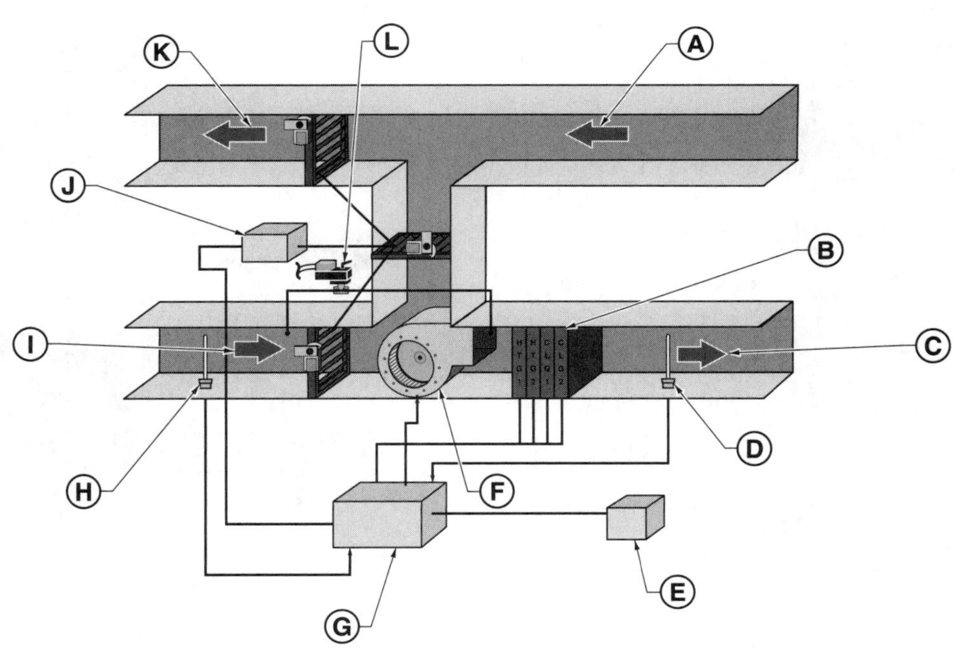

Pressure-Independent VAV Terminal Box with Reheat Coil

_______________ 1. primary air (55°F)

_______________ 2. room diffuser

_______________ 3. to controller

_______________ 4. reheat coil

_______________ 5. flow sensor at terminal box inlet

_______________ 6. VAV terminal box

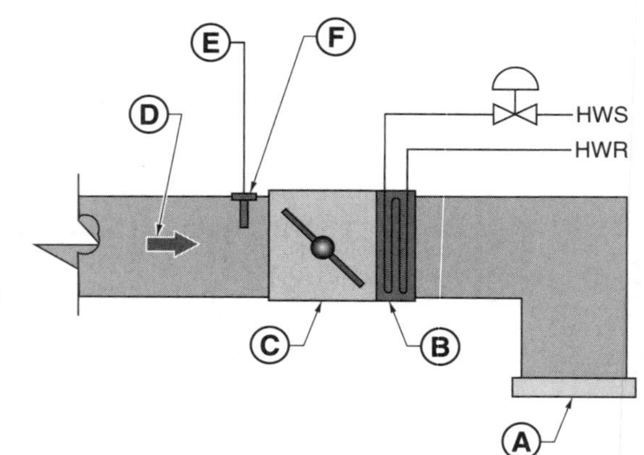

Minimum Static Pressure

_______________ 1. thermostat

_______________ 2. fan volume control

_______________ 3. filter

_______________ 4. static pressure sensor

_______________ 5. supply fan

_______________ 6. air handling unit controller

_______________ 7. return air

_______________ 8. mixed air plenum

_______________ 9. exhaust air

_______________ 10. volume damper

_______________ 11. cooling coil

_______________ 12. outside air

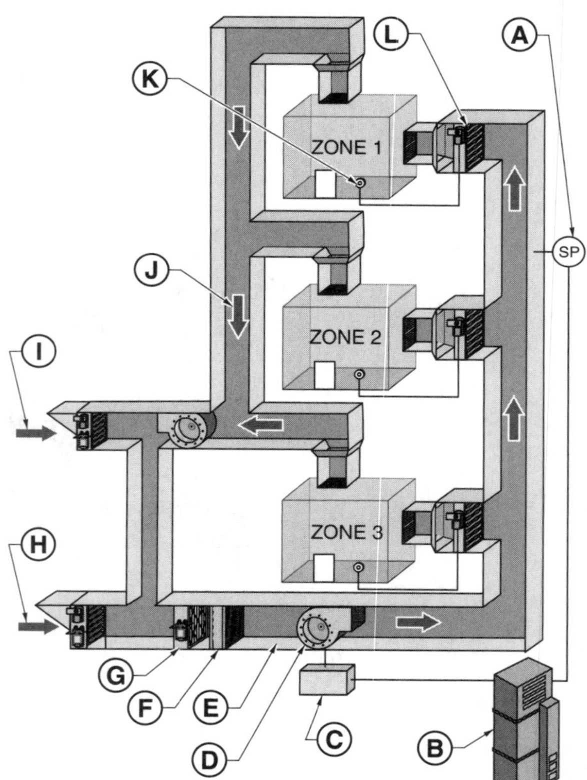

ACTIVITIES

Name ______________________________ **Date** ______________

Activity 17-1. Building Automation System Identification

Use the communication riser drawing to answer the questions.

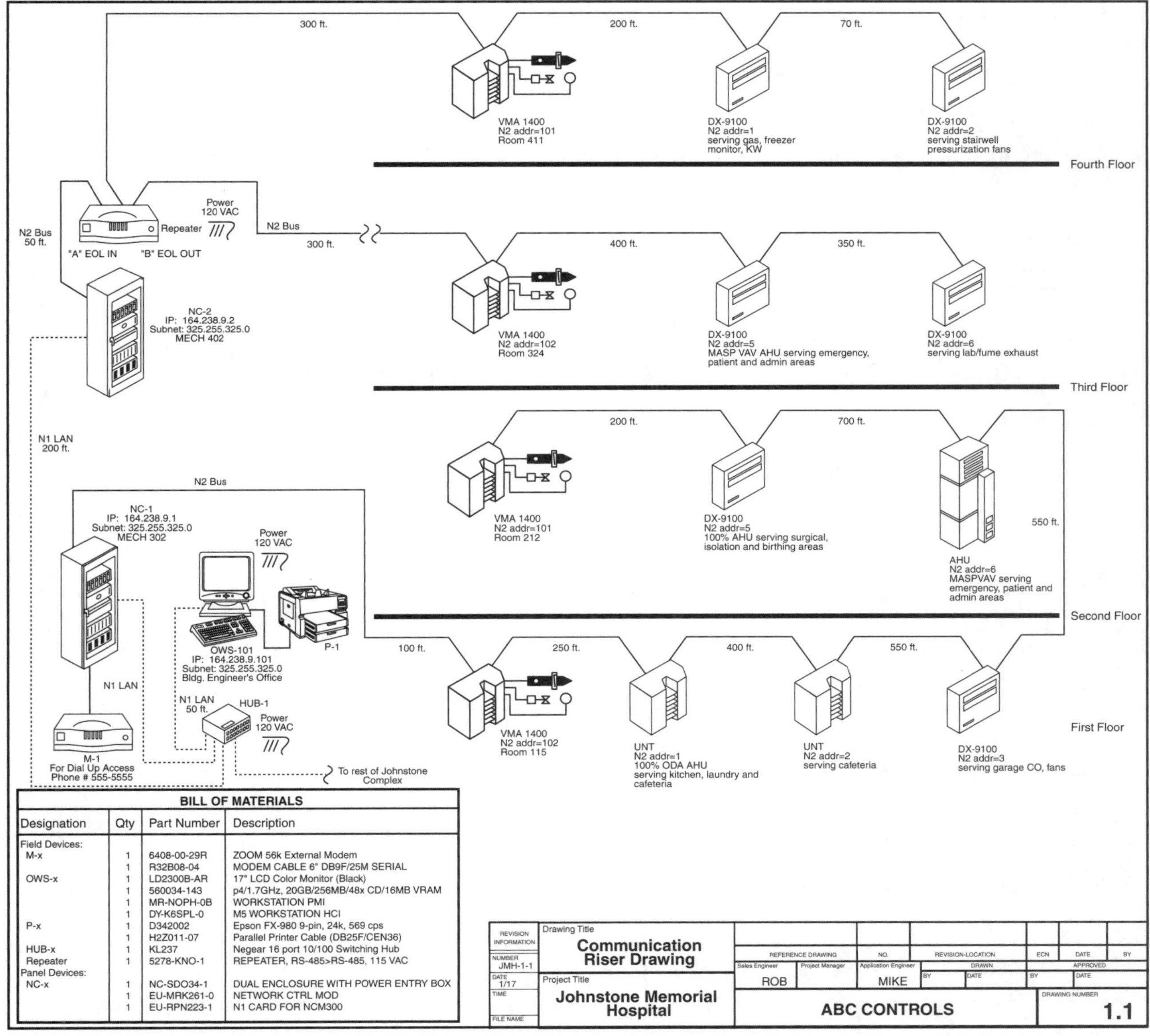

BILL OF MATERIALS			
Designation	Qty	Part Number	Description
Field Devices:			
M-x	1	6408-00-29R	ZOOM 56k External Modem
	1	R32B08-04	MODEM CABLE 6" DB9F/25M SERIAL
OWS-x	1	LD2300B-AR	17" LCD Color Monitor (Black)
	1	560034-143	p4/1.7GHz, 20GB/256MB/48x CD/16MB VRAM
	1	MR-NOPH-0B	WORKSTATION PMI
	1	DY-K6SPL-0	M5 WORKSTATION HCI
P-x	1	D342002	Epson FX-980 9-pin, 24k, 569 cps
	1	H2Z011-07	Parallel Printer Cable (DB25F/CEN36)
HUB-x	1	KL237	Negear 16 port 10/100 Switching Hub
Repeater	1	5278-KNO-1	REPEATER, RS-485>RS-485, 115 VAC
Panel Devices:			
NC-x	1	NC-SDO34-1	DUAL ENCLOSURE WITH POWER ENTRY BOX
	1	EU-MRK261-0	NETWORK CTRL MOD
	1	EU-RPN223-1	N1 CARD FOR NCM300

______________ **1.** Is the building automation system a central supervisory, central DDC, or distributed DDC system?

2. What items on the print might identify it as such?

____

____

______________ **3.** How many network communication modules are included in the system?

4. List each network communication module name and location.

____

____

5. What are the functions of the network communication module(s)?

____

____

______________ **6.** NC-2 is located a distance of ___′ from HUB-1.

______________ **7.** How many VAV terminal box controllers are included in the system?

8. List the location and local communication bus address for each VAV terminal box controller.

____

____

9. What type of programming would the VAV terminal box controllers use?

____

____

______________ **10.** In what kind of chip would their program be located?

______________ **11.** How many UIOMs are included in the system?

12. List the location, local communication bus address, and listed functions for each UIOM.

____

____

__________ **13.** Are the UIOMs easier or harder to program than the VAV terminal box controllers?

14. List the number, address, and HVAC unit served by the AHU controller(s).

____

____

An outside air temperature sensor is wired to the UIOM at address 3. The outside air temperature value must be shared with the AHU controller.

15. What type of data is this and what controller is responsible for sharing the data?

____

__________ **16.** The VAV terminal box controller in room 324 is located a distance of ___′ from the UIOM at address 5.

17. List the number, address, and function of all controllers located on the fourth floor.

____

____

__________ **18.** The hub part number is ___.

__________ **19.** The repeater part number is ___.

__________ **20.** The network control module part number is ___.

__________ **21.** The length of communication bus wire required for the job is ___′. *Note:* Add 10% for connections, wire stripping, etc.

Activity 17-2. Building Automation System Controller Specifications

Use the variable air volume box controller information on the following page to answer the questions.

__________ **1.** What type of controller is it?

2. List the controller power requirements.

____

3. List the controller primary control strategies.

____

4. List the controller model numbers that have screw terminal block connections.

____

5. Fan configurations that are available are ___.

____

6. List the number and type of analog inputs used with the AS-VAV1111-12 controller.

____

7. List the number and type of analog outputs used with the AS-VAV1111-12 controller.

____

______________ **8.** The required voltage and current of the binary outputs are ___.

Variable Air Volume Box Controller

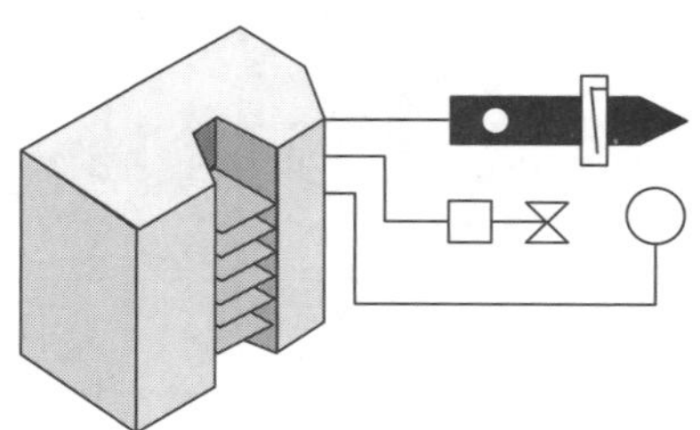

Description

The Variable Air Volume Box (VAV) Controller is specifically designed for digital control of single duct, dual duct, fan-powered, and supply/exhaust VAV box configurations. The controller can provide stand alone control of the VAV box, and can also integrate control of baseboard heat and lighting logic for the room or zone.

Features

- multiple modes of operation for various occupancy conditions
- stand alone control for small systems
- N2 bus communications and networking software capabilities
- interfaces to both electric and pneumatic actuators
- controller-resident performance calculations

OPTIONS	
Application	**Software**
Primary Equipment Types	VAX box, single duct, dual duct, fan powered or assisted, supply/exhaust
Primary Control Strategies	Pressure dependent, pressure independent, constant volume
Box Heat Configuration	Incremental, proportional, or two-position (NO or NC) valves 1-, 2-, or 3-stage electric
Baseboard Heat Configuration	Incremental, two-position, (NO or NC) valves, single stage electric
Cooling Configuration	Incremental output to damper actuator
Fan Configuration	Parallel, temperature setpoint; parallel, CFM setpoint; series, On-Off control; series, proportional control
Lighting Control	On and Off outputs to lighting relay in conjunction with occ/unocc mode
Unoccupied Control	Setup, setback, or shutdown

SPECIFICATIONS	
Variable Air Volume Box Controller	
Product	AS-VAV 1110-12, AS-VAV 1111-12, FA-VAV 1110-12, FA-VAV 1111-12 spade connector: AS-VAV 1140-12, AS-VAV 1141-12, FA-VAV 1140-12, FA-VAV 1141-12 screw terminal block
Ambient Operating Conditions	32°F to 140°F 10% to 90% RH
Dimensions (H × W × D)	6.5 × 6.4 × 4.0 in. 6.8 × 7.3 × 6.7 in. with enclosure
N2 Bus	Isolated
Zone Bus	8-pin phone jack or terminal block on controller
Power Requirements	24 VAC, 50/60 Hz, 10 VA plus binary output loads
Shipping Weight	2.6 lb

SELECTION CHART					
Model Number	**Termination Type**	**Analog Inputs**	**Binary Inputs**	**Analog Outputs**	**Binary Outputs**
AS-VAV1110-12 FA-VAV1110-12	Spade lug	6 RTD temperature element (NI, SI, or PT) 0-10 VDC transmitter 2 kΩ setpoint potentiometer	4 4-dry contacts 1 momentary pushbutton from zone sensor	—	8 24 VAC triacs at 0.5 A
AS-VAV1111-12 FA-VAV1111-12				2 0 VDC to 10 VDC at 10 mA	6 same as above
AS-VAV1140-12 FA-VAV1140-12	Screw terminal	6 RTD temperature element (NI, SI, or PT) 0-10 VDC transmitter 2 kΩ setpoint potentiometer	4 4-dry contacts 1 momentary pushbutton from zone sensor	—	8 24 VAC triacs at 0.5 A
AS-VAV1141-12 FA-VAV1141-12				2 0 VDC to 10 VDC at 10 mA	6 same as above

Section 18.1 On-Site Operator Interfaces

REVIEW QUESTIONS

Name ______________________________ **Date** ____________________

Multiple Choice

_______________ **1.** A(n) ___ is a device that allows an individual to access and respond to building automation system information.
A. controller
B. network communication module
C. termination device
D. operator interface

_______________ **2.** The primary desktop PC used to communicate with a building automation system is commonly referred to as the ___.
A. notebook
B. front end
C. dumb terminal
D. main

_______________ **3.** A(n) ___ is a controller-mounted device that consists of a small number of keys and a small display.
A. dumb terminal
B. portable operator terminal (POT)
C. keypad display
D. interface module

_______________ **4.** A keypad display can display ___ lines of information.
A. 2 to 4
B. 10 to 15
C. 20 to 60
D. 50 to 100

_______________ **5.** ___ are one-way operator interfaces with the building automation system.
A. Alarm printers
B. On-site desktop PCs
C. Keypad displays
D. Fax machines

_______________ **6.** Most building automation systems use ___ as the building automation system network.
A. nodes
B. Ethernet
C. virtual private networks (VPNs)
D. local area networks (LANs)

_______________ **7.** A ___ is a device that has a unique address and is attached to a network.
A. node
B. pager
C. supervisory controller
D. local area network (LAN)

_______________ **8.** Most modern building automation systems provide a(n) ___ at the temperature sensor for checking system operation.
A. keypad
B. thermostat
C. phone jack connection
D. overload protection device

_______________ **9.** An alarm printer may be connected to a ___.
A. network sensor
B. keypad display
C. field controller
D. desktop PC or a network communication module

_______________ **10.** The operator interface method most commonly used in commercial buildings is a(n) ___ connected to the building automation system network.
A. off-sit laptop PC
B. desktop personal computer (PC)
C. keypad display
D. smartphone

_______________ **11.** Keypad displays are commonly mounted in the ___.
A. 24-hour guard station
B. side panel of the field bus
C. boiler room office
D. front panel of a BAS network communication module

_______________ **12.** A ___ is a controller-mounted device that contains only a small number of keys and a small display.
A. keypad
B. laptop computer
C. desktop computer
D. node

Completion

_______________ **1.** ___ is a local area network (LAN) architecture that can connect up to 1024 nodes and supports data transfer rates of 10 megabits per second (Mbps).

_______________ **2.** A local area network (LAN) is a(n) ___ that spans a relatively small area.

_______________ **3.** A(n) ___ printer is a printer used with a building automation system to produce hard copies of alarms (indications of improper system operation), preventive maintenance messages, and data trends.

______________________ **4.** When loaded with the appropriate software, a(n) ___ can function as a portable service/maintenance tool that can be used to change setpoints and time schedules, or check a controller for proper operation.

______________________ **5.** A(n) ___ is a controller-mounted device that consists of a small number of keys and a small display.

______________________ **6.** A(n) ___ can display only two to four lines of information and can change only times and setpoints.

______________________ **7.** Most modern building automation systems provide a(n) ___ connection at the temperature sensor for checking system operation.

Building Automation System Network

_______________ 1. programmable controllers

_______________ 2. web browser

_______________ 3. supervisory controllers

_______________ 4. workstation

_______________ 5. network

_______________ 6. field controllers

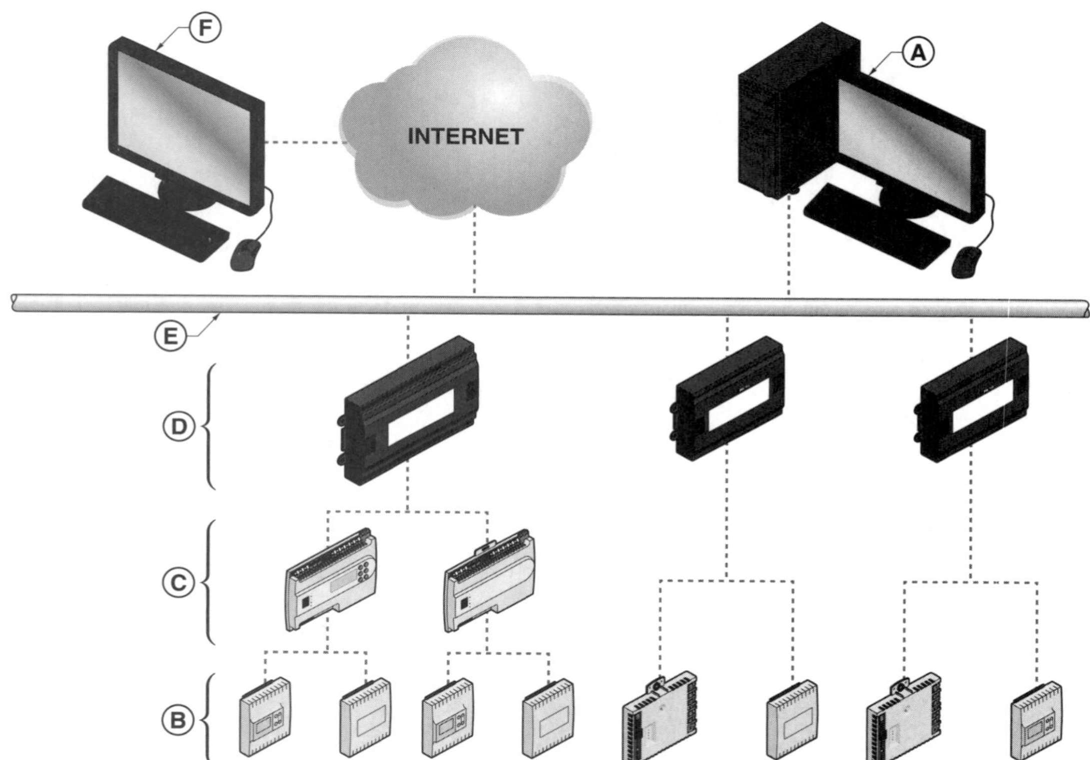

Section 18.2 Off-Site Operator Interfaces

REVIEW QUESTIONS

Name ______________________________ **Date** ________________

Multiple Choice

______________ **1.** A building automation system can send alarms and logs to ___ in addition to off-site PCs.
- A. pagers
- B. fax machines
- C. cell phones
- D. email

______________ **2.** The most common off-site operator interface is a(n) ___.
- A. off-site portable operator terminal
- B. fax machine
- C. off-site desktop PC
- D. smartphone

______________ **3.** The widespread use of ___ enables an alarm that occurs when a building is unoccupied to be reported to on-call maintenance personnel.
- A. private networks
- B. off-site alarm software
- C. personal computers and modems
- D. pagers

______________ **4.** A ___ may use a voice phone interface to enable changes to the setpoints and time schedules of a building automation system.
- A. cellular phone
- B. fax machine
- C. tablet
- D. pager

Completion

______________ **1.** A(n) ___ can be used to gain secure access into a business intranet and allow access to the building automation system.

______________ **2.** A(n) ___ is a small portable electronic device that vibrates, emits a beeping signal, or displays a text message when the individual carrying it is paged.

______________ **3.** Building automation software chooses the phone number of a person who is on call at the time the alarm occurs, enabling ___.

Section 18.3 Operator Interface Selection and Training

REVIEW QUESTIONS

Name ______________________________ **Date** ____________________

Multiple Choice

______________ **1.** One of the primary reasons for poor customer satisfaction with a building automation system is the ___.
   A. lack of effective training
   B. complexity of the system
   C. cost of the system
   D. time to install

______________ **2.** A(n) ___ is required if only a few individuals need access to the building automation system.
   A. laptop PC or smartphone
   B. basic operator interface
   C. portable operator terminal
   D. on-site desktop or laptop PC

______________ **3.** The ___ level access requires the highest amount of training.
   A. engineering
   B. supervisor
   C. lead technician
   D. technician

______________ **4.** The content of a training course should be examined closely and a training needs analysis survey should be performed on ___.
   A. a semiannual schedule
   B. each new building constructed
   C. each new employee
   D. each new building automation system installation

______________ **5.** The ___ is/are responsible for setting up policies and procedures regarding professional development goals, incentives for improvement, staffing needs, and long-term management goals.
   A. production department
   B. company owner(s)
   C. human resources department
   D. board of trustees

_______________ **6.** A(n) ___ is the group of items and functions that an operator is permitted to perform.
- A. access code
- B. access level
- C. technician level
- D. training clearance

Completion

_______________ **1.** The operator interface for a building automation system is selected according to the ___ of interface required.

_______________ **2.** A(n) ___ level is the group of items and functions that an operator is permitted to perform.

_______________ **3.** The ___ level allows the assigning of access levels and codes to other individuals.

_______________ **4.** Operators and technicians are issued a(n) ___ that activates the operator or technician access level.

_______________ **5.** When a(n) ___ is assigned, the operator is given an alphanumeric access code that is entered into the building automation system database.

_______________ **6.** A comprehensive training plan should include building automation system training as well as ___ and should be integrated with the human resources department of a facility.

Operator Interfaces

ACTIVITIES

Name ______________________________ **Date** ____________________

Activity 18-1. Operator Interface Identification

Use the communication riser drawing to answer the questions.

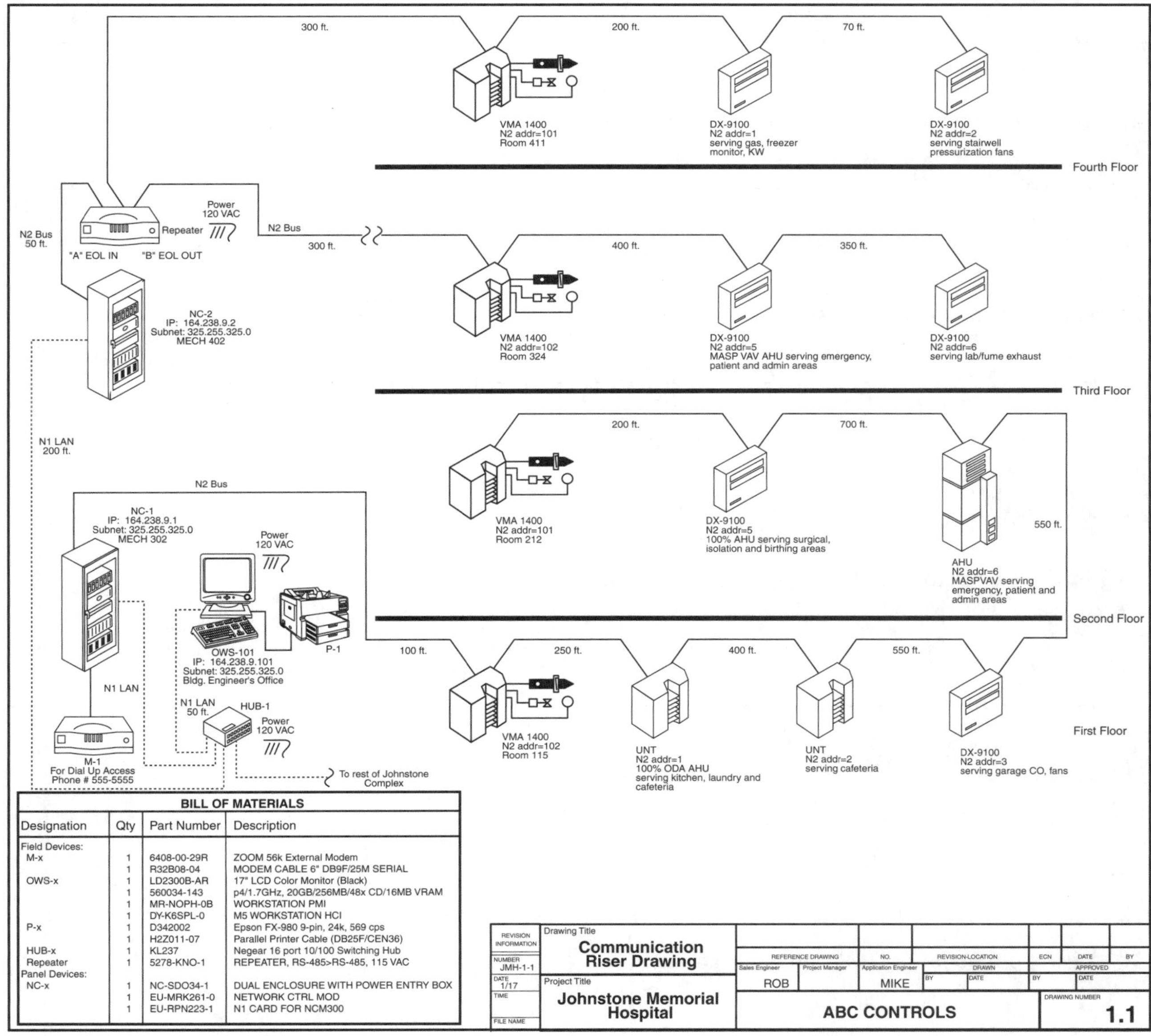

BILL OF MATERIALS

Designation	Qty	Part Number	Description
Field Devices:			
M-x	1	6408-00-29R	ZOOM 56k External Modem
	1	R32B08-04	MODEM CABLE 6" DB9F/25M SERIAL
OWS-x	1	LD2300B-AR	17" LCD Color Monitor (Black)
	1	560034-143	p4/1.7GHz, 20GB/256MB/48x CD/16MB VRAM
	1	MR-NOPH-0B	WORKSTATION PMI
	1	DY-K6SPL-0	M5 WORKSTATION HCI
P-x	1	D342002	Epson FX-980 9-pin, 24k, 569 cps
	1	H2Z011-07	Parallel Printer Cable (DB25F/CEN36)
HUB-x	1	KL237	Negear 16 port 10/100 Switching Hub
Repeater	1	5278-KNO-1	REPEATER, RS-485>RS-485, 115 VAC
Panel Devices:			
NC-x	1	NC-SDO34-1	DUAL ENCLOSURE WITH POWER ENTRY BOX
	1	EU-MRK261-0	NETWORK CTRL MOD
	1	EU-RPN223-1	N1 CARD FOR NCM300

1. List three operator interface devices shown on the drawing.

____

____

______________ **2.** What type of interface device is OWS-101?

3. Where is OWS-101 located?

____

____

4. List three functions that OWS-101 might perform.

____

____

5. What is the purpose of interface device M-1?

____

____

______________ **6.** What is its phone number?

7. List three operator interface devices that M-1 might report to.

____

____

______________ **8.** What type of operator interface device is P-1?

______________ **9.** What operator interface device is it connected to?

10. List two functions that P-1 might perform.

____

____

Activity 18-2. Using Operator Interfaces

Use operator interface A to answer the questions.

________________ **1.** What type of operator interface device is shown?

2. List several advantages of this type of operator interface.

____

____

____

OPERATOR INTERFACE A

Use operator interface B to answer the questions.

________________ **3.** What type of operator interface device is shown?

4. In what types of situations is this type of operator interface best used?

____

____

OPERATOR INTERFACE B

5. List all devices on the communication riser drawing that would use the same type of operator interface.

____

____

Use the supervisory controller drawing to answer the questions.

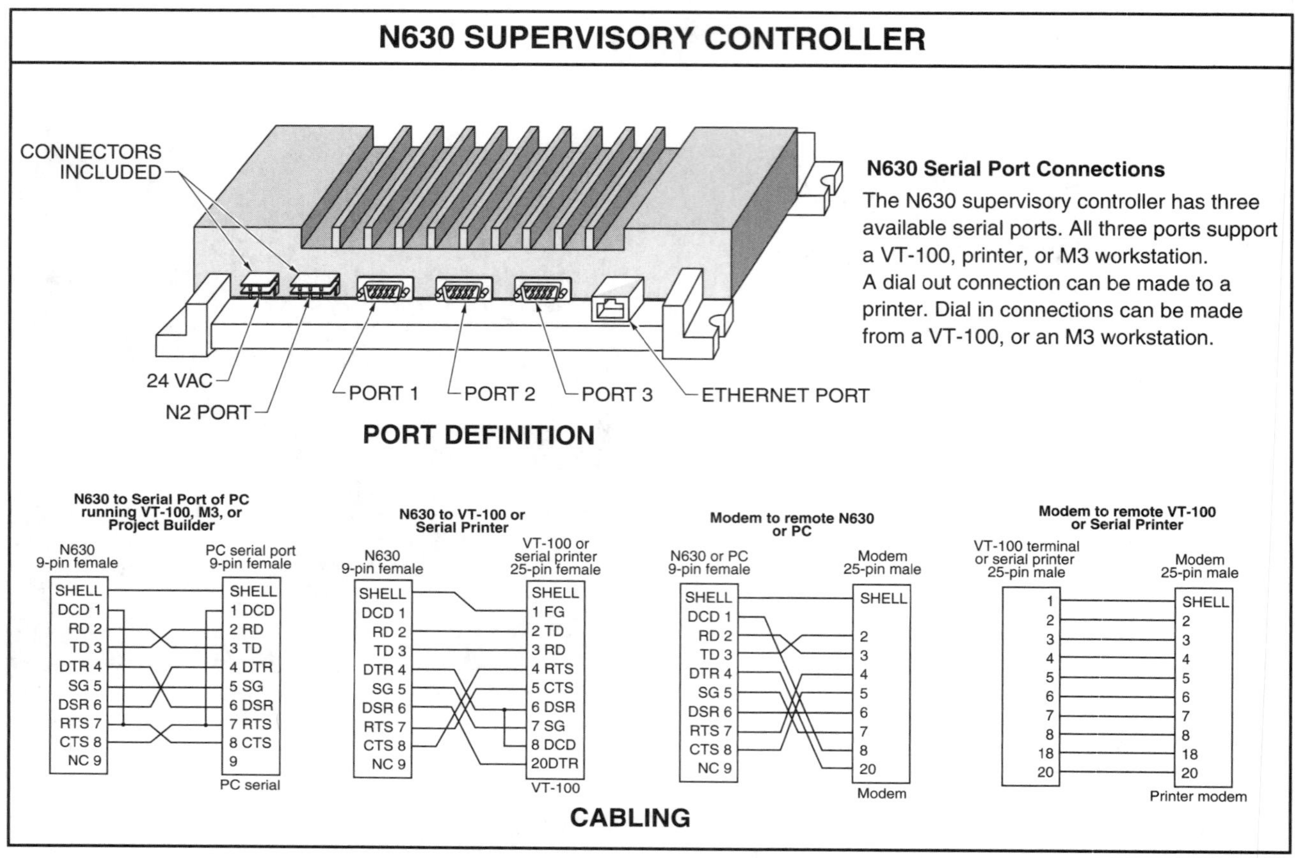

______________________ **6.** How many 9 pin serial ports are shown?

7. In the cabling details, list three of the operator interface devices shown.

____

A security guard must be able to receive a list of alarms from the supervisory controller.

8. What are two interface devices that might be used?

____

This controller is being installed at a hospital. The system requires that changes be made from a home office.

9. List three operator interface devices that enable off-site changes.

____

Activity 18-3. Training and Access Levels

A building automation system installation contract calls for quality factory-certified training to be provided.

1. What can be done to determine what type of training is required?

____

____

Use the training schedule and the communication riser drawing to answer the question.

2. List three training courses that may be appropriate for the system being installed in the Johnstone Memorial Hospital.

____

____

There are eight staff members in the building. Joe is the supervisor. Sam and Susan are the lead technicians. John, Sarah, Jerry, Dave, and Jim are shift operators. List the appropriate access levels (supervisor, engineer, lead technician, and technician) needed for each.

__________ **3.** Joe access levels are ___.

__________ **4.** Sam access levels are ___.

__________ **5.** Susan access levels are ___.

__________ **6.** John access levels are ___.

__________ **7.** Sarah access levels are ___.

__________ **8.** Jerry access levels are ___.

__________ **9.** Dave access levels are ___.

__________ **10.** Jim access levels are ___.

ABC CONTROLS COMPANY - Training Schedule

Subject to change; courses will be on an as-needed basis.

Course #	Course Name	Course Start	Course End	Location of Training
120	Facility Operators	January 21	January 22	Anderson College
130	Controller Engineering/ Programming	February 2	February 8	Anderson College
121	ASC Engineering	February 18	February 22	Anderson College
120	Facility Operators	March 18	March 22	Anderson College
127	Hardware Troubleshooting	April 8	April 12	Anderson College
167	Graphics Development	April 23	May 1	Anderson College
100	HVAC Basics	May 3	May 11	Anderson College
134	VAV Operators	May 21	May 22	Anderson College
120	Facility Operators	June 1	June 7	Anderson College
164	Control Engineering	June 17	June 21	Anderson College
100	HVAC Basics	July 5	July 11	Anderson College
124	HVAC Maintenance	July 22	July 26	Anderson College
122	LAN Engineering	August 2	August 3	Anderson College
120	Facility Operators	September 3	September 11	Anderson College
132	Software Update	September 24	September 28	Anderson College
121	ASC Engineering	October 14	October 18	Anderson College
100	HVAC Basics	November 2	November 7	Anderson College
131	Graphics Engineering	November 18	November 22	Anderson College

To enroll, please call 1-800-555-4343.

Activity 18-4. Adding Operator Interfaces

A customer wants to purchase operator interface devices that are not part of the original job. Use the communication riser drawing to answer the questions.

____________________ **1.** The system currently includes ___ (number) workstations.

____________________ **2.** Can a new workstation be added in Mech Room 401?

____________________ **3.** Can a printer be added?

____________________ **4.** If a printer is added, can it be any printer?

____________________ **5.** Can the network IP and subnet addresses be the same?

6. What suggestions can be made concerning suitable portable operator interface devices?

____

____

Section 19.1 Building Automation System Inputs

REVIEW QUESTIONS

Name ______________________________ **Date** ____________________

Multiple Choice

______________ **1.** A(n) ___ is a device that senses a variable such as temperature, pressure, or humidity and causes a proportional electrical signal change at the building automation system controller.

A. analog input
B. digital input
C. analog output
D. digital output

______________ **2.** A ___ temperature-coefficient temperature sensor increases its output resistance as the temperature increases, and decreases its output resistance as the temperature decreases.

A. negative
B. positive
C. neutral
D. reverse

______________ **3.** Humidity sensors drift over time, with inaccuracies of ___% per year being common.

A. 1
B. 2
C. 3
D. 5

______________ **4.** A ___ is a temperature-actuated switch.

A. digital input
B. timed override initiator
C. transducer
D. thermostat

______________ **5.** A(n) ___ is a device that records the number of occurrences of a signal.

A. current relay
B. accumulator
C. initiator
D. timer

______________________ **6.** A(n) ___ sensor is a humidity sensor that uses a thin film of hygroscopic element to alter the capacitance of a circuit.
- A. capacitive
- B. bulk polymer
- C. pressure
- D. enthalpy

______________________ **7.** ___ temperature sensors are the most common temperature sensors used in building automation systems and are used to sense air temperatures in building spaces.
- A. Wall-mount
- B. Duct-mount
- C. Well-mount
- D. Averaging

______________________ **8.** A ___ sends a digital input signal to a building automation system controller to indicate an alarm or fault condition.
- A. two-phase monitor
- B. card access controller
- C. timed override initiator
- D. flame safeguard controller

______________________ **9.** A(n) ___ is used to sense body heat when an individual enters a room.
- A. specialized digital input
- B. occupancy sensor
- C. timed override initiator
- D. split-core heat-sensing relay

______________________ **10.** Differential pressure switches are often used to indicate ___.
- A. the number of occurrences of a signal
- B. a difference in temperature across a coil or tube
- C. air leaks in switch tubing
- D. air handling unit filter condition

Completion

______________________ **1.** A(n) ___ is a resistor made of semiconductor material in which electrical resistance varies with changes in temperature.

______________________ **2.** All manufacturers provide ___ to prevent direct sunlight from affecting outside air temperature sensor operation.

______________________ **3.** A(n) ___ is the amount of light produced by a lamp (lumens) divided by the area that is illuminated.

______________________ **4.** A(n) ___ is a digital input device that switches open or closed because of the difference between two pressures.

______________________ **5.** A(n) ___ switch is a switch that contains a paddle that moves when contacted by air or water flow.

______________________ **6.** A(n) ___ is a device that, when closed, sends a signal to a controller which indicates that a timed override period is to begin.

______________________ **7.** A(n) ___ is a point that exists only in software and is not a hard-wired point.

______________________ **8.** A(n) ___ is a device that changes its characteristics as the humidity changes.

______________________ **9.** A(n) ___ is a sensor that detects whether an area is occupied by one or more individuals.

______________________ **10.** A building automation system ___ is a device that senses and sends building condition information to a controller.

______________________ **11.** A(n) ___ is a humidity sensor that consists of a polymer saturated with a salt compound.

______________________ **12.** A current-sensing relay is a device that surrounds a wire and detects the ___ due to electricity passing through the wire.

Flow Switches

______________________ **1.** paddle

______________________ **2.** water out

______________________ **3.** flow switch

______________________ **4.** pump

______________________ **5.** water in

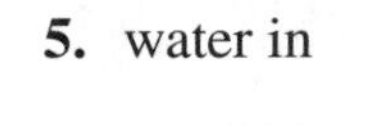

______________________ **6.** digital input

______________________ **7.** building automation system controller

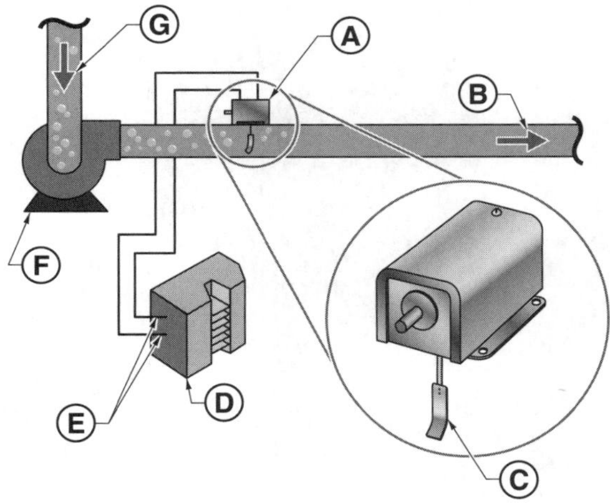

Temperature Sensor Mounting

____________________ **1.** well-mount

____________________ **2.** duct-mount

____________________ **3.** averaging

____________________ **4.** wall-mount

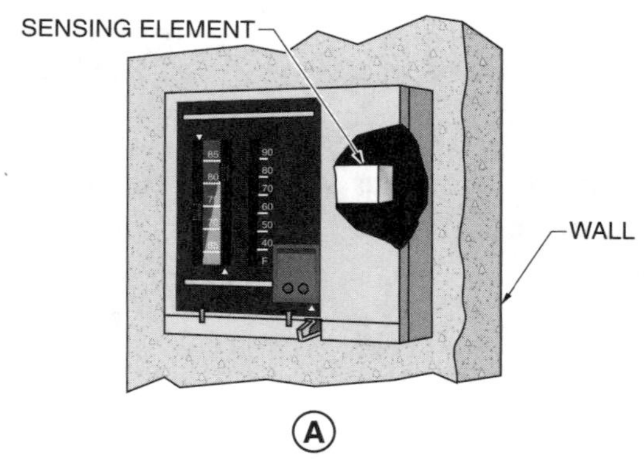

Ⓐ

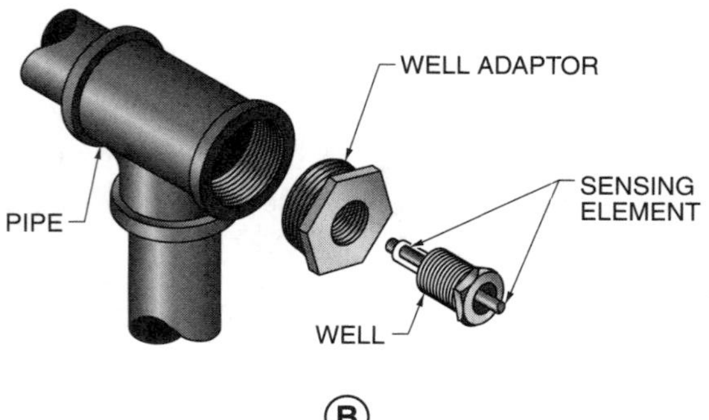

Ⓑ

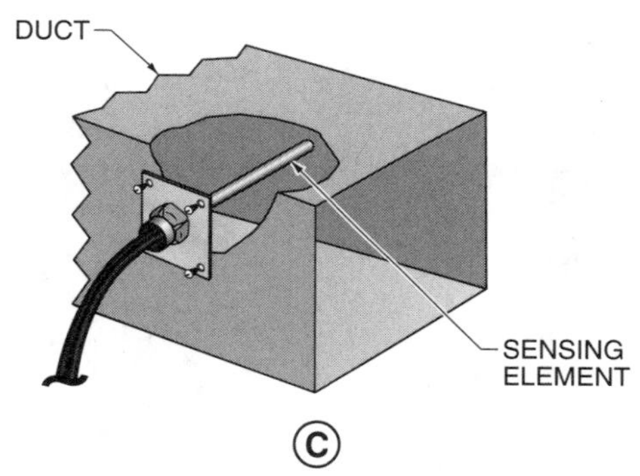

Ⓒ

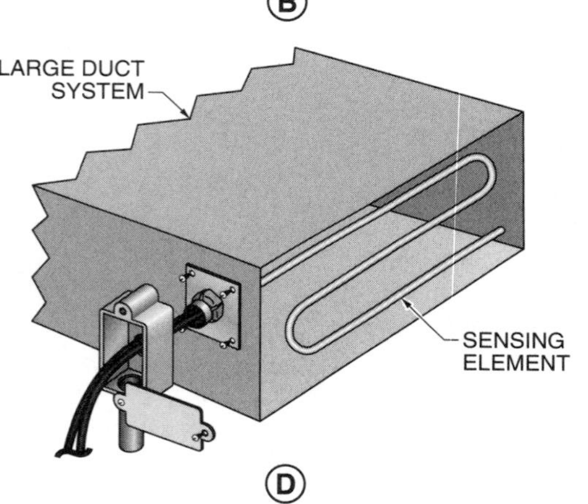

Ⓓ

Section 19.2 Building Automation System Outputs

REVIEW QUESTIONS

Name ______________________________ **Date** ______________________

Multiple Choice

______________ **1.** A ___ is a device that changes the state of a controlled device in response to a command from a building automation system controller.
A. valve or damper actuator
B. controller
C. triac
D. building automation system output

______________ **2.** ___ is a control technique in which a sequence of short pulses is used to position an actuator.
A. Gate turn ON
B. Latching
C. Pulse width modulation (PWM)
D. Bias voltage turn ON

______________ **3.** A(n) ___ is a device that produces a continuous signal between two values.
A. electronic input
B. solid-state input
C. digital output
D. analog output

______________ **4.** Transducers can receive input currents and voltages of 4 mA to 20 mA and 0 VDC to 10 VDC and output a voltage using a variable resistance such as 0 Ω to ___ Ω.
A. 85
B. 135
C. 210
D. 250

______________ **5.** A(n) ___ is a solid-state switching device used to switch alternating current.
A. triac
B. transducer
C. incremental
D. pulse width modulation

_______________ **6.** An incremental output is a digital output device used to position a(n) ___.
A. latching relay
B. triac
C. actuator
D. bidirectional electric motor

_______________ **7.** The ___ of the actuator is the length of time that it takes the actuator to move from one extreme to the other.
A. incremental output
B. stroke time
C. pulse width modulation
D. differential pressure

_______________ **8.** Pulse width modulation (PWM) is a control technique in which a sequence of short pulses is used to position a(n) ___.
A. transducer
B. output signal
C. actuator
D. latching relay

Completion

_______________ **1.** A(n) ___ is a device that accepts an ON or OFF signal.

_______________ **2.** A(n) ___ relay is a relay that requires a short pulse to energize the relay and turn ON the load.

_______________ **3.** A(n) ___ is a device that changes one type of proportional control signal into another.

_______________ **4.** A(n) ___ is a digital output device used to position a bidirectional electric motor.

ACTIVITIES

Name ______________________________ **Date** ______________

Activity 19-1. Analog Inputs

Use the panel board wiring diagram to answer the questions.

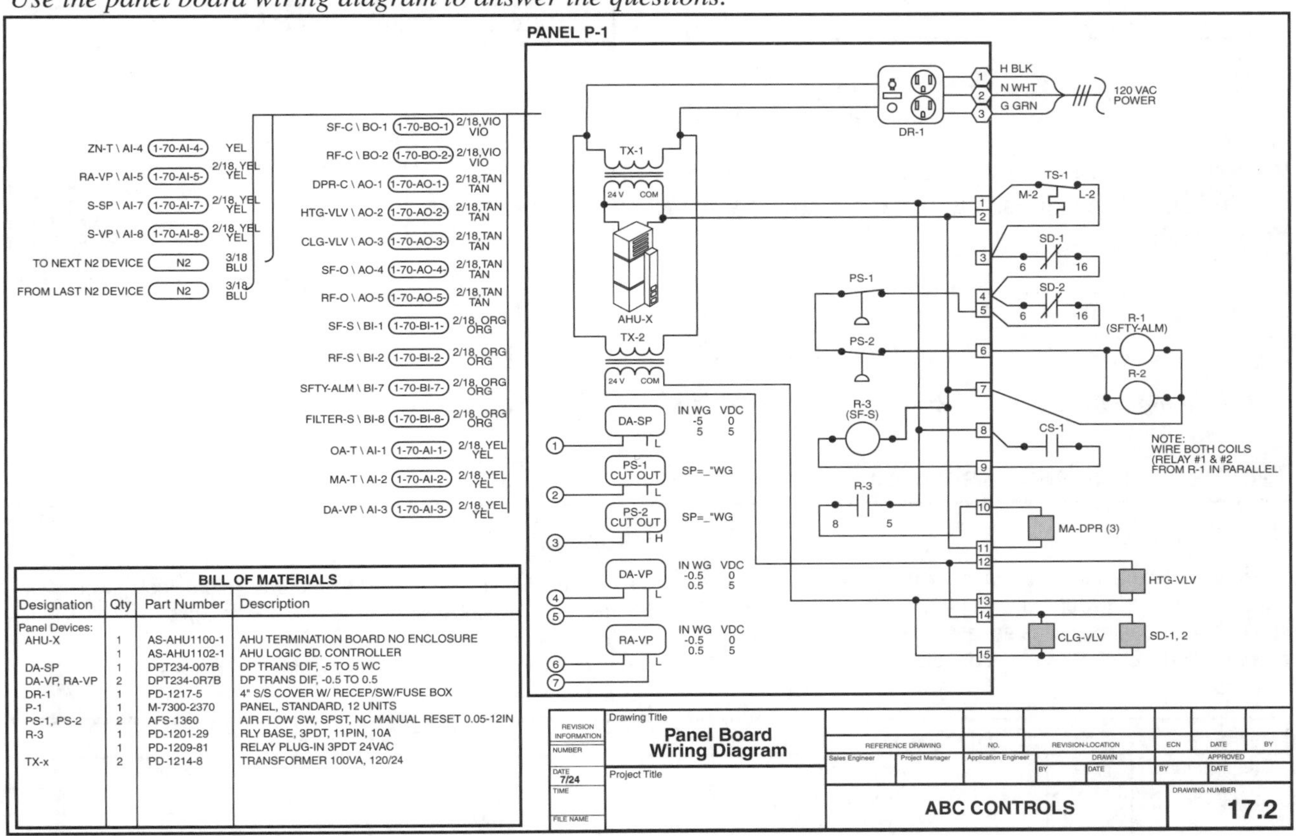

Designation	Qty	Part Number	Description
Panel Devices:			
AHU-X	1	AS-AHU1100-1	AHU TERMINATION BOARD NO ENCLOSURE
	1	AS-AHU1102-1	AHU LOGIC BD. CONTROLLER
DA-SP	1	DPT234-007B	DP TRANS DIF, -5 TO 5 WC
DA-VP, RA-VP	2	DPT234-0R7B	DP TRANS DIF, -0.5 TO 0.5
DR-1	1	PD-1217-5	4" S/S COVER W/ RECEP/SW/FUSE BOX
P-1	1	M-7300-2370	PANEL, STANDARD, 12 UNITS
PS-1, PS-2	2	AFS-1360	AIR FLOW SW, SPST, NC MANUAL RESET 0.05-12IN
R-3	1	PD-1201-29	RLY BASE, 3PDT, 11PIN, 10A
	1	PD-1209-81	RELAY PLUG-IN 3PDT 24VAC
TX-x	2	PD-1214-8	TRANSFORMER 100VA, 120/24

______________ **1.** How many analog inputs (AIs) are used in the system?

2. List three analog inputs.

____

______________ **3.** How many temperature sensors are used in the system?

______________ **4.** The pressure range of the DA-SP sensor is ___″ wc.

______________ **5.** Should the supply air velocity pressure sensor be located as close as possible to an elbow or tee to obtain a better reading?

Use drawing details B20 and B22 to answer the questions.

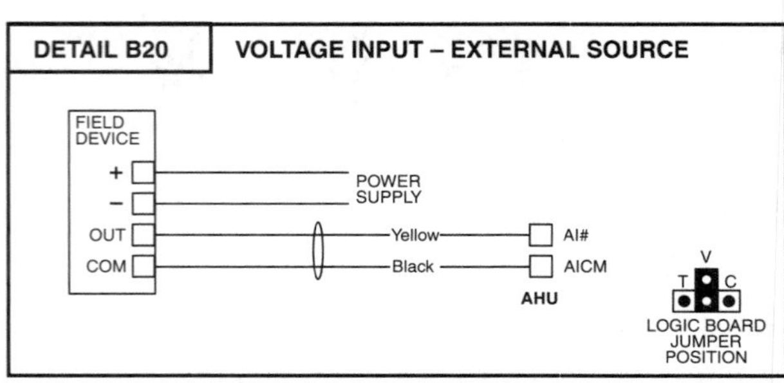

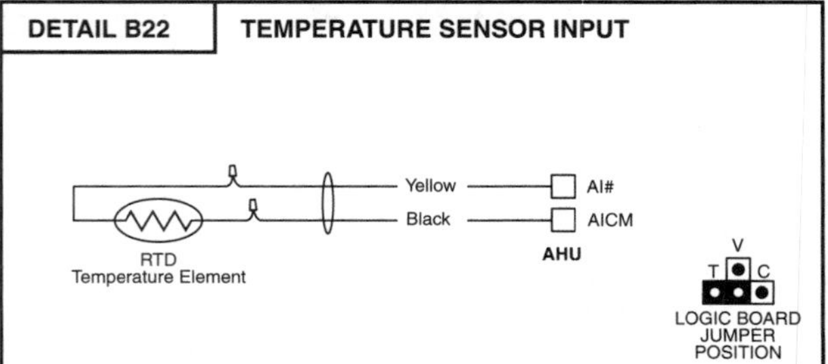

__________ **6.** What color wire(s) is/are used to connect the field device to the AHU controller?

7. What other wires that are needed for the sensor are indicated on detail B20?

____

____

__________ **8.** What type of analog input is indicated on detail B22?

9. Must the controller programming be consistent with the type of sensor used in the application? Why or why not?

____

____

Use the temperature sensor/assembly information to answer the questions.

Temperature/Sensor Assembly

Description

TEM-161 completed assemblies are used in a variety of temperature sensing applications. Various other sensing elements and hardware configurations are available that can be field-assembled, depending on the application.

Applications

- control or indication of high-temperature steam using well-insertion assemblies in hot water pipes or tanks
- temperature-averaging
- duct-insertion for controlling cycling in areas of sudden, large temperature changes

SELECTION CHART

Code Number	Type	Description
TEM-161-1	Nickel	17-ft averaging, temperature sensing element (1000 Ω, 1%) with handi-box
TEM-161-2	Nickel	Same as TEM-161-1 except that it has 8 ft averaging element
TEM-161-3	Nickel	High-temperature (550°F) well-insertion element (1000 Ω, 1%)
TEM-161-4	Nickel	Same as TEM-161-2 except with setpoint, without cover
TEM-161-54	Silicon	Base room thermostat w/setpoint, wo/cover
TEM-161-55	Silicon	Space temperature assembly with wall plate adaptor and mounting bracket

SPECIFICATIONS

TEM-161 Series Sensor/Hardware Assemblies

Elements	TEM-161-1 through -8	Nickel wire resistance type
	TEM-161-54, -55	PTC Silicon
Reference Resistances	TEM-161-1 through -8	1000 Ω at 70°F
	TEM-161-10	1000 Ω at 70°F, 50% RH
	TEM-161-54, -55	1035 Ω at 77°F
Temperature Coefficient	TEM-161-1 through -8	Positive, approximately 3 Ω/°F
	TEM-161-54, -55	Positive, approximately 4.3 Ω/°F
Tolerance Resistances	TEM-161-1, -2, -8	±1.0% at 70°F
	TEM-161-3	±1.0% at 70°F
	TEM-161-5	±1.0% at 70°F
	TEM-161-54, -55	±0.05% at 77°F
Set Point Range	TEM-161-8	55 to 85°F
	TEM-161-54	55 to 85°F

ACCESSORIES

Code Number	Description		
WZ-100-4	Stainless steel immersion well for use with TEM-161-3	TEM-161-62	Toggle switch for use with TEM-161-54, -55, -4
TEM-161-61	Pushbutton switch for use with TEM-161-54, -55, -4	TEM-180-96	Electrical wall box mounting adapter kit includes wallplate adapter, mounting bracket, and screws

__________ **10.** If an 8′ averaging element sensor is required for an application, which sensor should be selected?

__________ **11.** Which sensor should be selected for a base room thermostat without a cover but with a setpoint?

__________ **12.** Which type of sensor is it?

__________ **13.** If a pushbutton accessory is needed for a room thermostat, what number should be selected?

__________ **14.** In the specification chart, what is the sensor type for a TEM 161-54 sensor?

__________ **15.** What is the reference resistance of the TEM 161-54 sensor?

__________ **16.** What is the temperature coefficient of the TEM 161-54 sensor?

__________ **17.** What is the resistor tolerance of the TEM 161-54 sensor?

Activity 19-2. Digital Inputs

Use the panel board wiring diagram on page 159 to answer the questions.

__________ **1.** How many digital (binary) inputs are used in the system?

2. List three digital inputs.

____

____

__________ **3.** Are any electric meters indicated?

4. What is the purpose of filter-S?

____

____

__________ **5.** The part number for PS-1 is ___.

__________ **6.** The part number for PS-2 is ___.

__________ **7.** What type of reset do PS-1 and PS-2 have?

8. Why do they have this type of reset?

____

____

Use the hot water temperature control information to answer the questions.

Hot Water Temperature Control

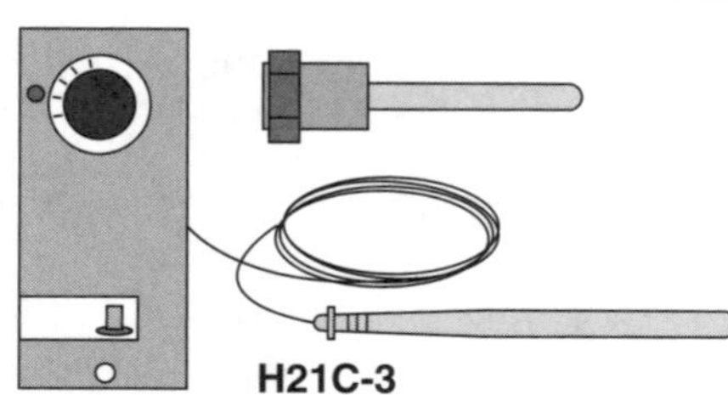

Description

This is a universal replacement for open high or SPDT applications. The control is furnished with a well assembly for ½″ tapping.

Applications

- operating control for hot water boilers

Features

- liquid-filled element provides rapid response to temperature change
- adjustable differential

SELECTION CHART							
Code Number	**Application**	**Switch Action**	**Range***	**Differential***	**Well Connection Size—NPT**	**Range Adjuster**	**Maximum Bulb Temperature***
H21C-1	Open High (R-B) Open Low (R-Y)	SPDT	100 to 240	6 to 24	½ in.	Convertible	250
H21C-2		SPDT	100 to 240	6 to 24	½ in. 8 ft Cap.	Convertible	290
H21D-3	High Temp Lockout	Open High with Lockout	100 to 240	Manual Reset (locks out high)	½ in.	Knob	250

* in °F

__________ **9.** The code number of the control that requires a manual reset is ___.

__________ **10.** This control's maximum bulb temperature is ___°F.

__________ **11.** Is a separate well assembly required, or is it provided?

__________ **12.** The switch action of control H21C-2 is ___.

__________ **13.** On control H21C-2, terminals ___ open on a drop in temperature.

__________ **14.** Control H21C-2 range is ___°F.

__________ **15.** Control H21C-2 differential is ___°F.

Activity 19-3. Analog Outputs

Use the panel board wiring diagram on page 159 to answer the questions.

__________ **1.** How many analog outputs are used in the system?

2. List three analog outputs.

____

3. List the wire colors that are shown connected to the AHU controller.

____

Use the rooftop air handling unit drawing to answer the questions.

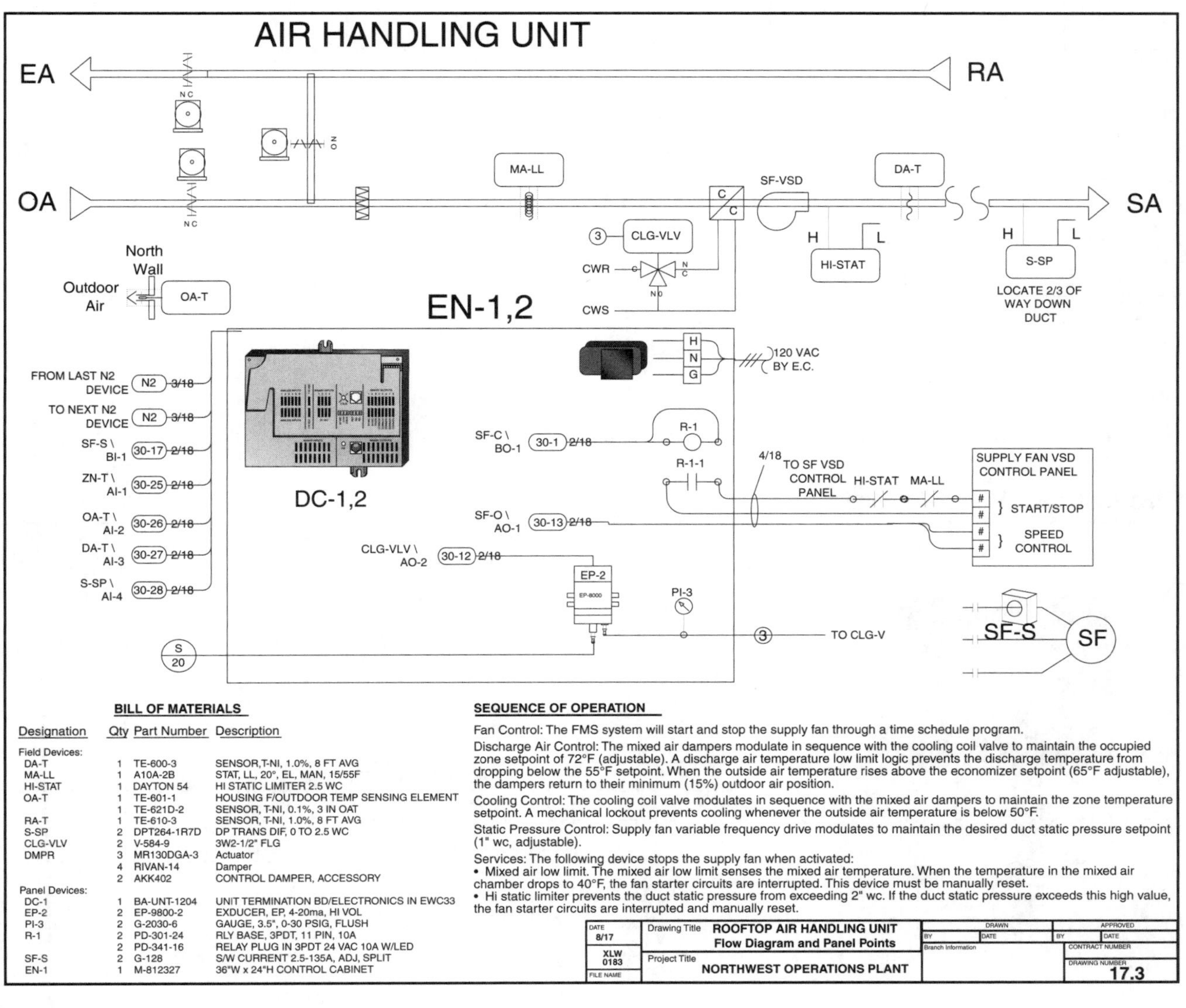

BILL OF MATERIALS

Designation	Qty	Part Number	Description
Field Devices:			
DA-T	1	TE-600-3	SENSOR,T-NI, 1.0%, 8 FT AVG
MA-LL	1	A10A-2B	STAT, LL, 20°, EL, MAN, 15/55F
HI-STAT	1	DAYTON 54	HI STATIC LIMITER 2.5 WC
OA-T	1	TE-601-1	HOUSING F/OUTDOOR TEMP SENSING ELEMENT
	1	TE-621D-2	SENSOR, T-NI, 0.1%, 3 IN OAT
RA-T	1	TE-610-3	SENSOR, T-NI, 1.0%, 8 FT AVG
S-SP	2	DPT264-1R7D	DP TRANS DIF, 0 TO 2.5 WC
CLG-VLV	2	V-584-9	3W2-1/2" FLG
DMPR	3	MR130DGA-3	Actuator
	4	RIVAN-14	Damper
	2	AKK402	CONTROL DAMPER, ACCESSORY
Panel Devices:			
DC-1	1	BA-UNT-1204	UNIT TERMINATION BD/ELECTRONICS IN EWC33
EP-2	2	EP-9800-2	EXDUCER, EP, 4-20ma, HI VOL
PI-3	2	G-2030-6	GAUGE, 3.5", 0-30 PSIG, FLUSH
R-1	2	PD-301-24	RLY BASE, 3PDT, 11 PIN, 10A
	2	PD-341-16	RELAY PLUG IN 3PDT 24 VAC 10A W/LED
SF-S	2	G-128	S/W CURRENT 2.5-135A, ADJ, SPLIT
EN-1	1	M-812327	36"W x 24"H CONTROL CABINET

SEQUENCE OF OPERATION

Fan Control: The FMS system will start and stop the supply fan through a time schedule program.

Discharge Air Control: The mixed air dampers modulate in sequence with the cooling coil valve to maintain the occupied zone setpoint of 72°F (adjustable). A discharge air temperature low limit logic prevents the discharge temperature from dropping below the 55°F setpoint. When the outside air temperature rises above the economizer setpoint (65°F adjustable), the dampers return to their minimum (15%) outdoor air position.

Cooling Control: The cooling coil valve modulates in sequence with the mixed air dampers to maintain the zone temperature setpoint. A mechanical lockout prevents cooling whenever the outside air temperature is below 50°F.

Static Pressure Control: Supply fan variable frequency drive modulates to maintain the desired duct static pressure setpoint (1" wc, adjustable).

Services: The following device stops the supply fan when activated:
- Mixed air low limit. The mixed air low limit senses the mixed air temperature. When the temperature in the mixed air chamber drops to 40°F, the fan starter circuits are interrupted. This device must be manually reset.
- Hi static limiter prevents the duct static pressure from exceeding 2" wc. If the duct static pressure exceeds this high value, the fan starter circuits are interrupted and manually reset.

DATE 8/17	Drawing Title **ROOFTOP AIR HANDLING UNIT Flow Diagram and Panel Points**	DRAWN BY / DATE	APPROVED BY / DATE
XLW 0183	Project Title **NORTHWEST OPERATIONS PLANT**	Branch Information	CONTRACT NUMBER
FILE NAME			DRAWING NUMBER 17.3

4. What does the SF-O do?

____

____

5. What is EP-2?

____

____

______________ **6.** What device is EP-2 connected to?

______________ **7.** Would the cooling valve have a spring range?

8. How do you know?

____

____

9. If only one analog output is used for three damper actuators, what does this indicate?

____

____

______________ **10.** Must the controller programming be consistent with the type of sensor used in the application?

Use the proportional valve actuator information to answer the questions.

Proportional Valve Actuator

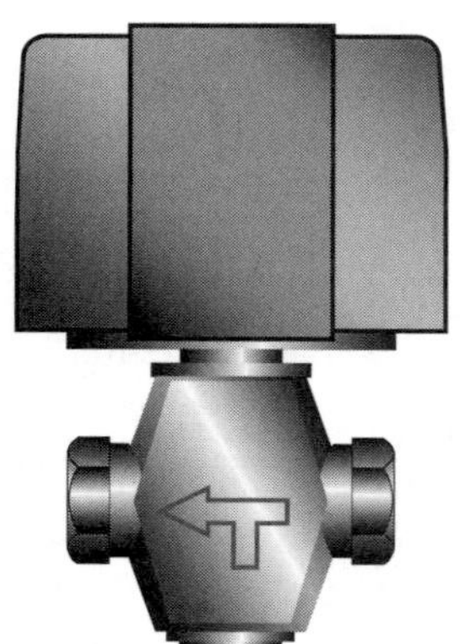

Description

The PVA-122 proportional valve actuator is an electric actuator that provides proportional control valves with up to a 5⁄16″ stroke in HVAC applications.

Features

- simplified setup and adjust procedures decrease installation costs
- wide range of control inputs meets the needs of most applications
- Light Emitting Diode (LED) reduces commissioning time and displays operating status

SPECIFICATIONS		
Power Requirements	24 VAC at 50/60 Hz, 5.0 VA supply minimum, Class 2	
Input Signal	Jumper Selectable: Factory Setting:	0 to 10 VDC, 0 to 20 VDC, 6 to 9 VDC, 0 to 20 mA 0 to 10 VDC, Drive Down action on signal increase
Input Impedance	Voltage Input: Current Input:	0 to 10 VDC, 150,000 Ω and 0 to 20 VDC, 450,000 Ω 0 to 20 mA, 500 Ω
Force	Shutoff and Breakaway: 22 lb minimum	
Cycles	60,000 full cycles; 1,500,000 repositions	
Enclosure	NEMA 1	
Ambient Operating Conditions	35 to 135°F Maximum dew point at 90% RH, non-condensing	

______________ **11.** What is the factory setting for the input signal?

______________ **12.** What are the actuator power requirements?

13. What input signals may be selected with a jumper?

____

____

______________ **14.** What is the shutoff force?

______________ **15.** What displays the operating status of the controller?

Activity 19-4. Digital Outputs

Use the panel board wiring diagram on page 159 to answer the questions.

____________________ **1.** How many binary outputs are used in the system?

2. List the binary outputs.

____

3. A number of binary outputs are connected to an output relay. Why?

____

____

____________________ **4.** Is any lighting control indicated?

____________________ **5.** Are any incremental actuators indicated?

____________________ **6.** How many binary outputs are used per incremental output?

Use the electric valve actuator information to answer the questions.

Electric Valve Actuator

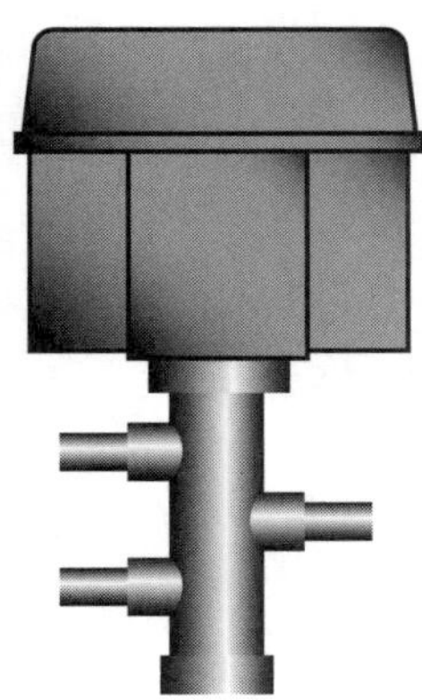

Description

The EVA-850 synchronous, motor-driven actuator provides floating control of valves with up to ¾″ stroke in HVAC applications. This compact, non-spring return actuator has a 50 lb force and requires a 3-wire, 24 VAC signal from the controller.

SPECIFICATIONS	
Product	EVA-850-1 Electric Valve Actuator Assemblies
Control Mode	Floating Control, 3-Wire
Supply Voltage	24 VAC +6 V, –4 V, <200 mA, 50/60 Hz
Power Consumption	6 VA
Shutoff Force	50 lb Force Minimum
Stroke Time	1/2 in. Stroke: Approx. 65 Sec 3/4 in. Stroke: Approx. 90 Sec
Ambient Operating Temperature	0 to 140°F, 10 to 90% RH Non-Condensing, 85°F Max Dew Point
Media Temperature	Water: 190°F Steam: 280°F

Operation

A controller sends 24 VAC to the up or down terminal on the circuit board depending on the desired movement of the valve. This signal causes the motor to rotate in the proper direction, moving the valve up or down. When the controller stops sending a signal, the valve remains in place.

When the controller closes the valve, a shutoff force builds. When this force reaches 50 lb, the lever activates a force sensor which stops the motor.

____________________ **7.** The maximum stroke for the actuator is ___″.

____________________ **8.** Does the actuator have a spring return?

____________________ **9.** What type of control does it support?

10. What happens to the valve when the controller stops sending a signal?

____

____

____________________ **11.** The actuator power consumption is ___ VA.

____________________ **12.** The approximate stroke time for a ½″ stroke valve is ___ sec.

Section 20.1 Building Automation System Installation

REVIEW QUESTIONS

Name ______________________________ **Date** ____________________

Multiple Choice

______________ **1.** All building automation system controllers should be wired in accordance with the ___ and local regulations.
A. Occupational Safety and Health Administration (OSHA)
B. American National Standards Institute (ANSI)
C. National Electrical Code® (NEC®)
D. International Electrotechnical Commission (IEC)

______________ **2.** Operator interfaces commonly require installation ___″ above the finished floor for viewing by maintenance personnel.
A. 48
B. 60
C. 72
D. 84

______________ **3.** A DIN rail is a flat mounting rail that is attached to a ___.
A. mounting point
B. control panel
C. battery
D. power supply

______________ **4.** A common building automation system controller mounting requirement includes separate building automation system controller conductors and power conductors by at least ___′.
A. 1
B. 2
C. 3
D. 5

______________ **5.** According to the NEC®, a Class 2 power-limited transformer is limited to ___ VA to reduce the shock hazard to humans.
A. 45
B. 50
C. 75
D. 100

_______________ **6.** A(n) ___ loop is a circuit that has more than one point connected to earth ground, with a voltage potential difference between the two ground points high enough to produce a circulating current in the ground system.
A. open
B. closed
C. ground
D. resistance

_______________ **7.** Facilities with critical systems use ___ for controllers in important building systems to allow the facilities to continue to operate during brief power failures.
A. uninterruptible power supplies
B. step-up transformers
C. backup batteries
D. isolated power supplies

_______________ **8.** The most common building automation system controller power supply is a ___ step-down transformer.
A. 120 VAC to 24 VAC
B. 190 VAC to 100 VAC
C. 230 VAC to 120 VAC
D. 320 VAC to 210 VAC

Completion

_______________ **1.** A(n) ___ is a flat rail attached to a control panel and used for mounting.

_______________ **2.** A(n) ___ is a circuit that has more than one point connected to earth ground, with a voltage potential difference between the two ground points high enough to produce a circulating current in the ground system.

_______________ **3.** Correct transformer ___ is essential within building automation systems because controller power supplies, sensors, and outputs are designed to operate properly when connected correctly.

Section 20.2 Building Automation System Wiring

REVIEW QUESTIONS

Name ______________________ **Date** ______________________

Multiple Choice

______________ **1.** ___ is a multipoint communication standard that incorporates low-impedance drivers and receivers providing high tolerance to noise.
A. RS-485
B. Dip switch
C. Peripheral wiring
D. Interface cabling

______________ **2.** Optoisolation is a communication method in which controllers use ___ components to prevent communication problems.
A. series
B. parallel
C. photonic
D. incremental

______________ **3.** Controller analog output terminals are commonly labeled ___, and COM.
A. V1, V2, etc.
B. T1, T2, etc.
C. DO 1, DO 2, etc.
D. AO 1, AO 2, etc.

______________ **4.** A(n) ___ is a device that is connected to a personal computer or building automation system controller to perform a specific function.
A. incremental output device
B. jumper
C. normally open contact
D. peripheral device

______________ **5.** A(n) ___ network configuration is a control network configuration in which controllers are connected to the network that runs throughout a building through spliced drops for each controller.
A. daisy chain
B. star
C. multidrop
D. Ethernet loop

______________ **6.** A ___ is a solid-state switching device used to switch alternating current.
A. conductor
B. ground loop
C. triac
D. controller

Completion

______________ **1.** ___ terminations are resistors that act as an electronic damper to prevent unwanted signal reflections at the end of the wire.

______________ **2.** A(n) ___ network configuration is a configuration in which multiple controllers are connected in series.

______________ **3.** Digital inputs are commonly configured as ___.

______________ **4.** In optoisolated communication, the binary (0 or 1) information transmitted is converted to a(n) ___ that is processed by the controller.

______________ **5.** ___ is a communication protocol that uses small, low-power digital devices.

______________ **6.** A(n) ___ network is a special variation on an RS-485 control network.

______________ **7.** A(n) ___ is a unique number assigned to each building automation system controller on a communication network.

______________ **8.** ZigBee is commonly used in industrial equipment that requires ___-range wireless data transfer at relatively low rates.

______________ **9.** ___ reduces the effect of electromagnetic interference on communication network wiring.

Building Automation System Network Configuration

______________ **1.** multidrop

______________ **2.** star

______________ **3.** daisy chain

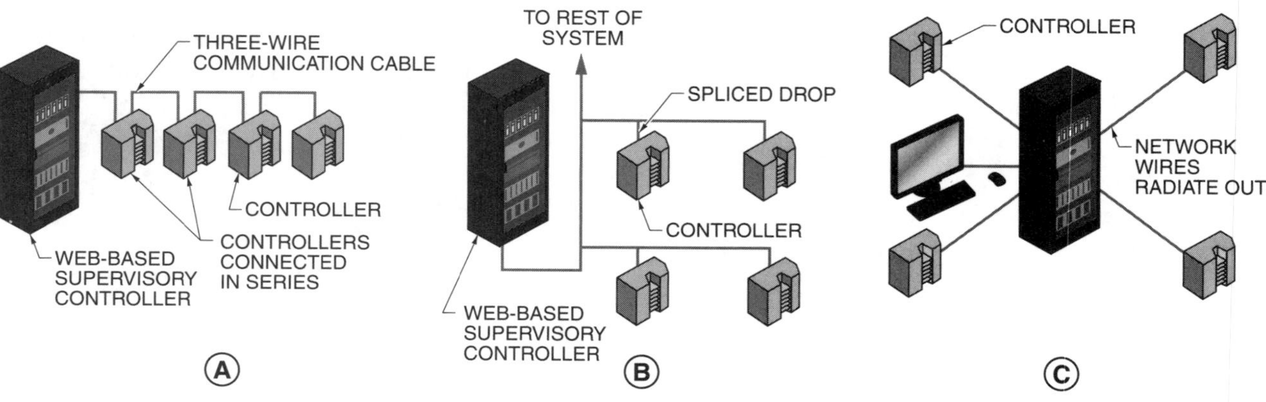

Section 20.3 Building Automation System Testing and Diagnostic Procedures

REVIEW QUESTIONS

Name ________________________________ Date ____________________

Multiple Choice

______________ **1.** When the voltage values of the RS-485+ and RS-485– conductors are measured, the voltage should be between ___.
- A. 3 VDC and 5 VDC
- B. 15 VAC and 30 VAC
- C. 1 mV and 10 mV
- D. 4 mV and 20 mV

______________ **2.** A common thermistor nominal value is 1000 Ω at ___°F.
- A. 50
- B. 70
- C. 90
- D. 110

______________ **3.** A ___ is used to test a controller power supply to ground.
- A. digital multimeter
- B. peripheral wiring device
- C. DIN rail
- D. jumper

______________ **4.** ___ testing determines whether there is a complete circuit between the building automation system controller and digital input/output devices.
- A. Digital input and output
- B. Continuity
- C. Network isolation
- D. Resistance

______________ **5.** ___ are used on building automation system controllers to select the type of analog input or output signal received by the controller.
- A. Digital multimeters
- B. Transmitters
- C. Jumpers
- D. Peripheral devices

Completion

______________ **1.** The resistance of a negative temperature coefficient sensor ___ as the temperature increases.

______________ **2.** The ___ of a thermistor sensor is the amount of resistance change per degree Fahrenheit.

______________ **3.** A(n) ___ is a conductor used to connect pins on a controller or device.

Controller Jumper Usage

______________ **1.** analog output voltage selection jumper

______________ **2.** analog output current selection jumper

______________ **3.** analog input current selection jumper

______________ **4.** jumper placement location

______________ **5.** analog input voltage/resistance selection jumper

______________ **6.** rear of controller

______________ **7.** jumper selection label

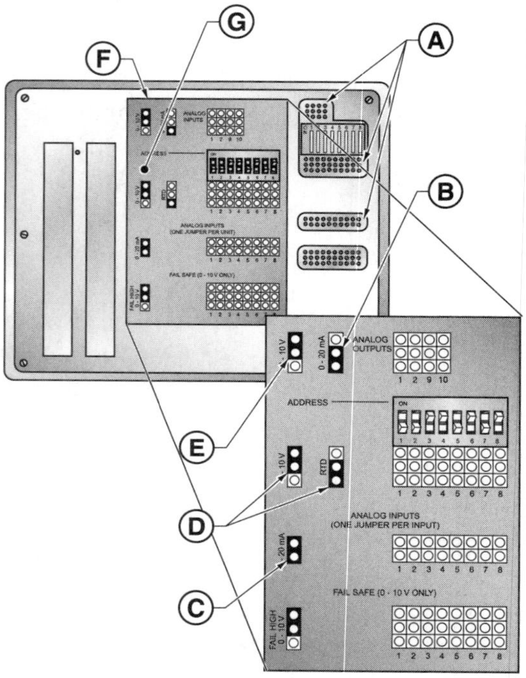

Building Automation System Installation, Wiring, and Testing

ACTIVITIES

Name ____ **Date** ______________________

Activity 20-1. Communication Bus, Peripherals, and Power Wiring

Use the building automation system wiring legend to answer the questions.

BUILDING AUTOMATION SYSTEM WIRING LEGEND							
CABLE/WIRE SPECIFICATION			TERMINALS PER CONDUCTOR COLOR				
Usage	**Part Number**	**Description**	**Black**	**White**	**Jacket Color**	**Red**	
AI	CBL-18/2YEL	18/2 Shld Yellow	AI Com	———	AI	———	
AI	CBL-18/3YEL	18/3 Shld Yellow	AI Com	Power	AI	———	
AO	CBL-18/2TAN	18/2 Shld Tan	AO Com	———	AO	———	
AO	CBL-18/3TAN	18/3 Shld Tan	AO Com	Power	AO	———	
BI	CBL-18/2ORG	18/2 Shld Orange	BI 24V	———	BI	———	
BO	CBL-18/2VLT	18/2 Shld Violet	BO Com	———	BO	———	
BO	CBL-18/3VLT	18/3 Shld Violet	BO Com	BO	BO	———	
GENERAL PURPOSE	CBL-18/2NAT	18/2 Shld Natural	Common	———	———	Power	
GENERAL PURPOSE	CBL-18/3NAT	18/3 Shld Natural	Common	———	Signal	Power	
CONTROLLER	CBL-24/8NAT	24/8 Natural	———	———	———	———	
CONTROLLER PHONE JACK	CBL-STAT25	Pre-Term'd Blue	———	———	———	———	
24 VAC	CBL-18/2GRY	18/2 Shld Grey	24 V Com	———	24 VAC	———	
24 VAC POWER BUS	PB0137	14/2 Unsh White	Common	———	———	24 VAC	
600 V	CBL-18/2600	18/2	———	———	———	———	
600 V	CBL-18/3600	18/3	———	———	———	———	
N2 BUS	CBL-18/3BLU	18/3 Shld Blue	N2-	N2-	N2+	———	
N1 BUS-ARCNET	CBL-RG62PUR	RG62 Purple	———	———	———	———	
N1 BUS-ETHERNET	63609	24/8 Purple	———	———	———	———	
BACNET BUS	63609	24/8 Purple	———	———	———	———	
LON BUS	43701	22/2 Shld Blue	———	———	———	———	
NT BUS	00-4340	22/4 Shld Blue	———	———	———	———	
XT BUS	CBL-18/3BLU	18/3 Shld Blue	XT-	XT-	XT+	———	

ETHERNET PATCH CABLE	
Pin No.	**Color**
1	WHT/ORN
2	ORN/WHT
3	WHT/GRN
4	BLU/WHT
5	WHT/BLU
6	GRN/WHT
7	WHT/BRN
8	BRN/WHT

ETHERNET CROSSOVER CABLE	
Pin No.	**Color**
1	WHT/GRN
2	GRN/WHT
3	WHT/ORG
4	BLU/WHT
5	WHT/BLU
6	ORN/WHT
7	WHT/BRN
8	BRN/WHT

1. What is the N1 bus Ethernet description?

____

____

____________________ **2.** What is the 24 VAC power bus part number?

____________________ **3.** For an Ethernet patch cable, what pin number is associated with the WHT/GRN colors?

Activity 20-2. Input Identification

Use detail B20 and B22 to answer the questions.

1. List the wiring terminals and wire colors.

____

____

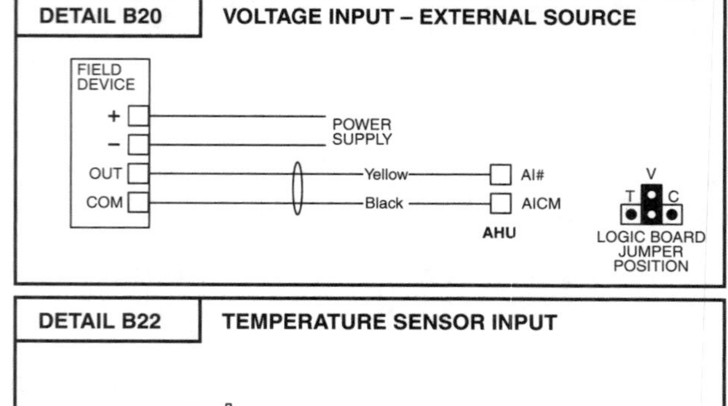

________________ **2.** If the resistance increases as the temperature increases, what kind of sensor is it?

________________ **3.** Are any jumpers involved for the setup of this AI?

________________ **4.** In detail B20, is the power supply polarity indicated?

Activity 20-3. Binary Inputs

Use the panel board wiring diagram to answer the questions.

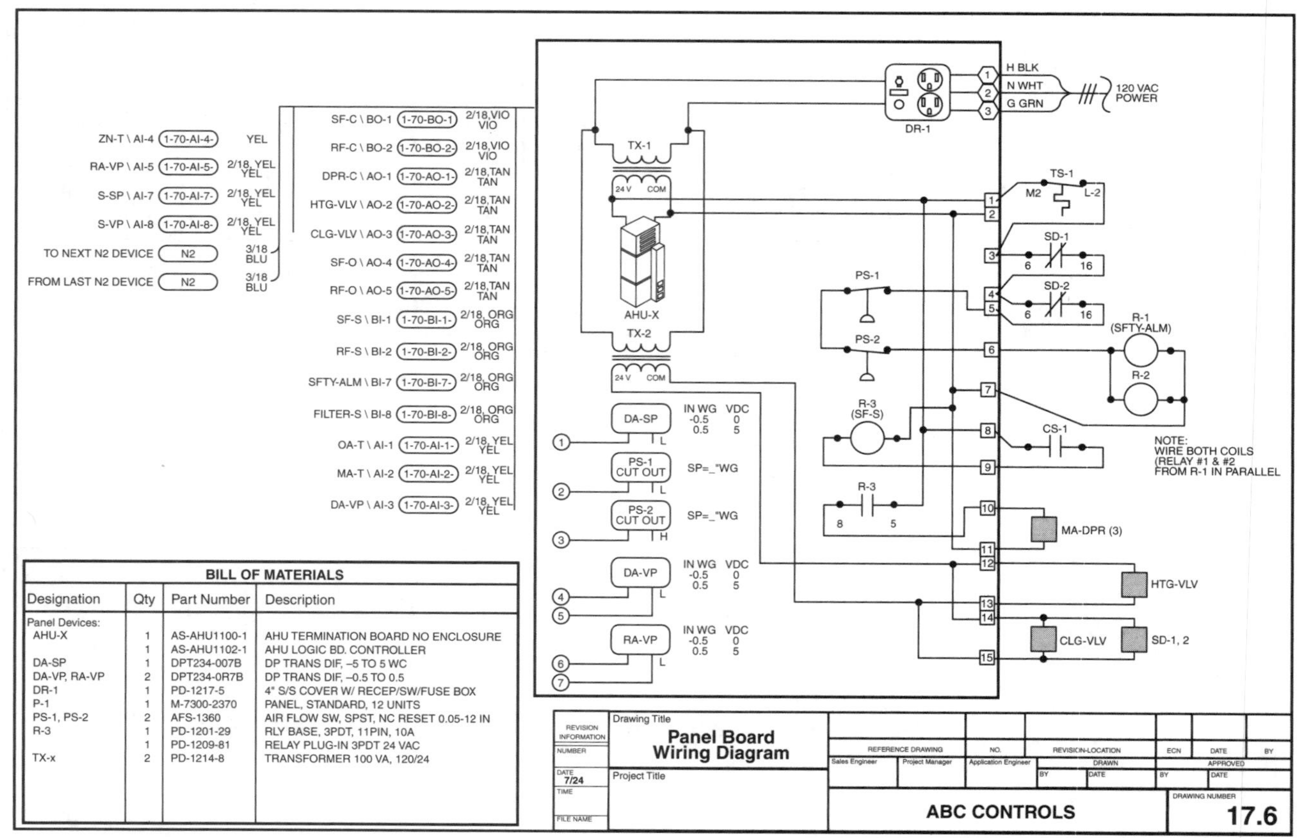

BILL OF MATERIALS

Designation	Qty	Part Number	Description
Panel Devices:			
AHU-X	1	AS-AHU1100-1	AHU TERMINATION BOARD NO ENCLOSURE
	1	AS-AHU1102-1	AHU LOGIC BD. CONTROLLER
DA-SP	1	DPT234-007B	DP TRANS DIF, –5 TO 5 WC
DA-VP, RA-VP	2	DPT234-0R7B	DP TRANS DIF, –0.5 TO 0.5
DR-1	1	PD-1217-5	4" S/S COVER W/ RECEP/SW/FUSE BOX
P-1	1	M-7300-2370	PANEL, STANDARD, 12 UNITS
PS-1, PS-2	2	AFS-1360	AIR FLOW SW, SPST, NC RESET 0.05-12 IN
R-3	1	PD-1201-29	RLY BASE, 3PDT, 11PIN, 10A
	1	PD-1209-81	RELAY PLUG-IN 3PDT 24 VAC
TX-x	2	PD-1214-8	TRANSFORMER 100 VA, 120/24

________________ **1.** To which binary input is the filter-S wired?

2. List the number of wires, wire gauge, and colors used for this binary input.

____

Use the building automation system wiring legend on page 173 to answer the question.

3. List the part number(s) and description(s) of all binary input cable.

____

Activity 20-4. Binary Outputs

Use the panel board wiring diagram on page 174 to answer the questions.

________________ **1.** To which binary output is RF-C wired?

2. List the number of wires, wire gauge, and colors used for this binary output.

____

Use the building automation system wiring legend on page 173 to answer the question.

3. List the part number(s) and description(s) of all binary output cable.

____

____

Activity 20-5. Analog Outputs

Use the panel board wiring diagram on page 174 to answer the question.

1. List the analog outputs, their use, and wire colors.

____

____

____

Use detail B24 to answer the questions.

__________________ **2.** What wire color(s) is/are used at the analog outputs?

__________________ **3.** Where are they terminated?

__________________ **4.** Is the polarity of the field device given?

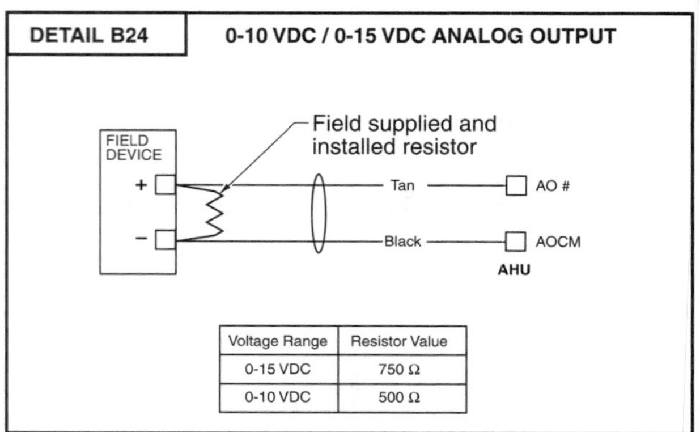

Use the building automation system wiring legend on page 173 to answer the question.

5. List the part number(s) and description(s) of all analog output cable.

____

____

Activity 20-6. Temperature Sensor Resistance/Temperature Characteristics

Use the temperature sensor resistance/temperature characteristic charts to answer the questions.

__________________ **1.** The approximate resistance value for a platinum sensor at a temperature of 80°F is ____ Ω.

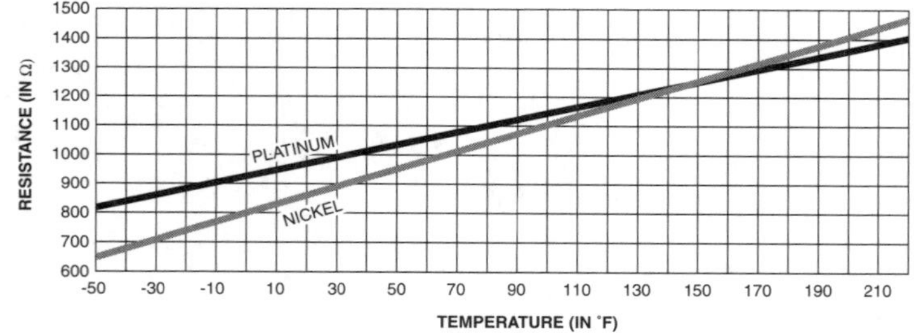

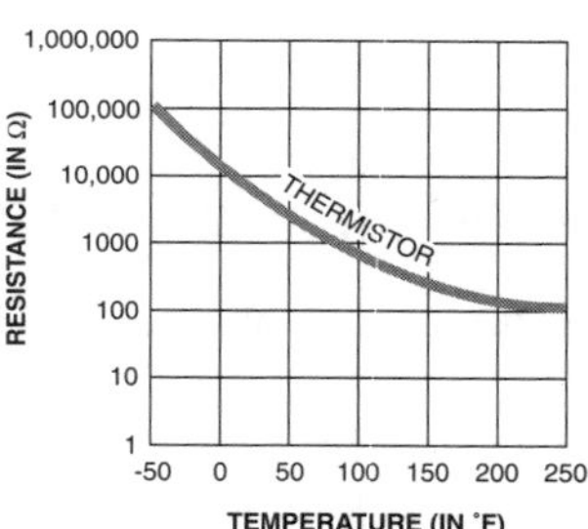

__________________ **2.** The approximate resistance value for a nickel sensor at a temperature of 50°F is ____ Ω.

__________________ **3.** The approximate resistance value for a thermistor sensor at a temperature of 150°F is ____ Ω.

Activity 20-7. Point Identification

Identify each point on the air handling unit as an analog input, analog output, binary input, or binary output.

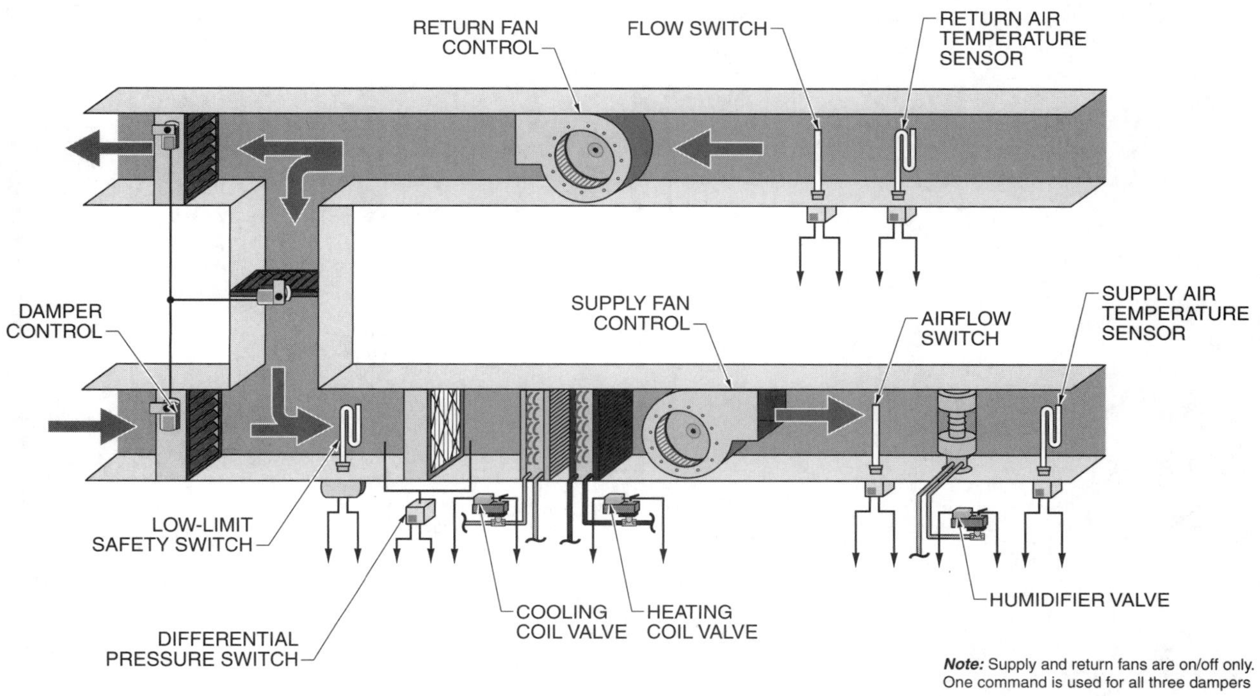

_____________________ **1.** Return fan control

_____________________ **2.** Flow switch

_____________________ **3.** Return air temperature sensor

_____________________ **4.** Supply air temperature sensor

_____________________ **5.** Humidifier valve

_____________________ **6.** Airflow switch

_____________________ **7.** Supply fan control

_____________________ **8.** Heating coil valve

_____________________ **9.** Cooling coil valve

_____________________ **10.** Differential pressure switch

_____________________ **11.** Low-limit safety switch

_____________________ **12.** Damper control

Section 21.1 Networks

REVIEW QUESTIONS

Name ______________________________ **Date** ____________________

Multiple Choice

______________ **1.** A ___ is a device, such as a computer or a printer, that has a unique address and is attached to a network.
A. node
B. topology
C. router
D. thinnet

______________ **2.** Ethernet is a local area network architecture that can connect up to ___ of nodes and supports minimum data transfer rates of 10 megabits per second (Mbps).
A. tens
B. hundreds
C. thousands
D. millions

______________ **3.** In an Ethernet network, data is divided into ___ before being transmitted.
A. receipt notices
B. concentrators
C. bundles
D. packets

______________ **4.** A switch manages the communication between networks or parts of a network that operate at ___ data transmission speeds.
A. low
B. high
C. different
D. the same

______________ **5.** ___ is the cables and other network devices, such as hubs and switches, that make up a network.
A. Hub
B. Thicknet
C. Physical layer
D. Ping

______________ **6.** The most common personal computer operating system used today is ___.
A. Microsoft Windows
B. Ethernet
C. Macintosh[SM]
D. OS10

______________ **7.** A(n) ___ is an address that can change at any time.
A. static address
B. dynamic address
C. mixed network address
D. intranet

______________ **8.** The ___ command displays the computer name on the network.
A. ipconfig
B. ping
C. hostname
D. MAC

______________ **9.** A(n) ___ network is a network in which new Ethernet network components are added to older obsolete network components.
A. intranet
B. mixed
C. Bluetooth
D. thicknet

______________ **10.** The Bluetooth device operating range depends on the device class and can range from 1′ up to ___′.
A. 10
B. 50
C. 150
D. 300

______________ **11.** A ___ allows a device to simultaneously communicate with a maximum of seven other devices.
A. thicknet
B. thinnet
C. piconet
D. personal area network (PAN)

______________ **12.** A(n) ___ is used to pass information across long distances involving multiple geographical locations and tens of thousands of users.
A. local area network (LAN)
B. wide area network (WAN)
C. node
D. Ethernet

____________ **13.** A ___ is a concentrator that manages the communication between different networks.
A. router
B. hub
C. firewall
D. subnetwork

Completion

____________ **1.** A(n) ___ normally encompasses one business or building and does not require the linking of a large number of users in remote locations.

____________ **2.** Network ___ is the map of the network configuration.

____________ **3.** A(n) ___ occurs when two nodes transmit simultaneously and data is corrupted.

____________ **4.** A(n) ___ is a network switchboard that allows a number of nodes to communicate with each other.

____________ **5.** A(n) ___ manages the communication between different networks.

____________ **6.** A(n) ___ cable is a cable used for connections from node to node.

____________ **7.** ___ is thick coaxial cable classified as 10base5.

____________ **8.** ___ addressing is a method of assigning addresses to nodes on a network.

____________ **9.** A(n) ___ is an echo message and its reply sent by one network device to detect the presence of another device.

____________ **10.** A(n) ___ address is hard-coded into the device by a manufacturer.

____________ **11.** A(n) ___ acts as a central device that connects the computers to each other and to the network or the Internet.

____________ **12.** ___ is a graphical option of the ipconfig command.

____________ **13.** ___ is a data transmission method where data is sent without frequency modulation.

____________ **14.** Bluetooth technology is used to allow short-range communication between devices for checkout and ___.

____________ **15.** ___ is an addressing scheme that filters messages and determines if the message is to be passed to local nodes on a subnetwork or if it is to be sent on to the main network through the router.

____________ **16.** A(n) ___ is a network switchboard that allows a number of nodes to communicate with each other.

____________ **17.** Each device on the Internet has an address known as the ___ address.

____________ **18.** A(n) ___ is a card installed in a network component to allow it to communicate with the network.

____________ **19.** ___ is thin coaxial cable classified as 10base2.

Ethernet Network Star Configurations

_______________ **1.** local communication bus

_______________ **2.** operator workstation

_______________ **3.** controller

_______________ **4.** central device (network communication module)

_______________ **5.** Ethernet local area network

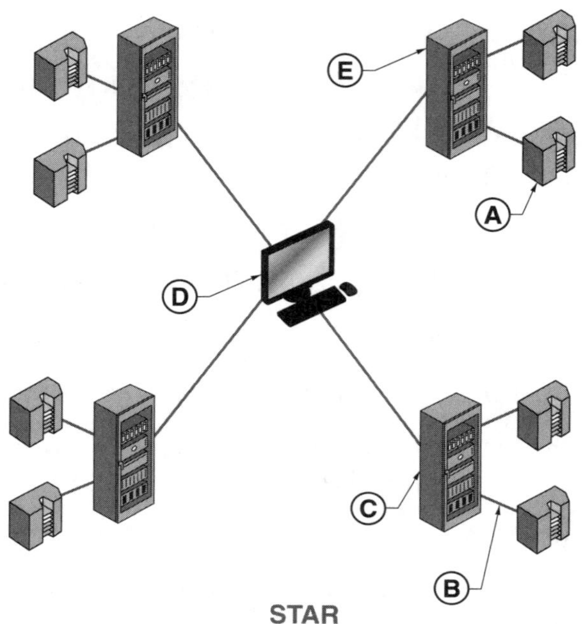

Section 21.2 Web-Based Control Systems

REVIEW QUESTIONS

Name ______________________ **Date** ______________________

Multiple Choice

______________ **1.** ___ software may be installed to allow remote access to the facility network using encryption and security techniques.
A. Virtual private network (VPN)
B. Personal area network (PAN)
C. Wired equivalent privacy (WEP)
D. Media access control (MAC)

______________ **2.** The ___ is the current concept and ability to add almost any building equipment and function type to a network.
A. Structured query language
B. mobile network
C. Internet of Things
D. remote desktop

______________ **3.** ___ are often used in a BAS to connect two devices without using a switch or router.
A. Hubs
B. Crossover cables
C. Dynamic addresses
D. Patch cables

______________ **4.** When the entire operating system of the device has to be replaced, it is known as "___" the device.
A. flashing
B. wiping
C. imagining
D. cleaning

______________ **5.** Disabling ___ may be necessary since they can prevent proper operation.
A. JAVA
B. popup blockers
C. firewalls
D. security programs

Completion

______ 1. Security for web-enabled control systems is provided by the network ___, which does not allow unauthorized access.

______ 2. The ___ is put in the address line of a browser.

______ 3. __ is a programming language designed for managing data in large database systems.

______ 4. Any PC that supports a(n) ___ can be used to access a web-based control system.

______ 5. Because of security concerns, computer and network ___ have becomes longer and more complex.

______ 6. A Windows feature called ___ can be used to move (update) software to web-based devices.

ACTIVITIES

Name ______________________ **Date** ______________

Activity 21-1. Building Automation System Troubleshooting

A building automation system does not appear to be communicating properly on the network. Before the manufacturer is called for service, basic tests are run on the system. Use the communication riser drawing to answer the questions.

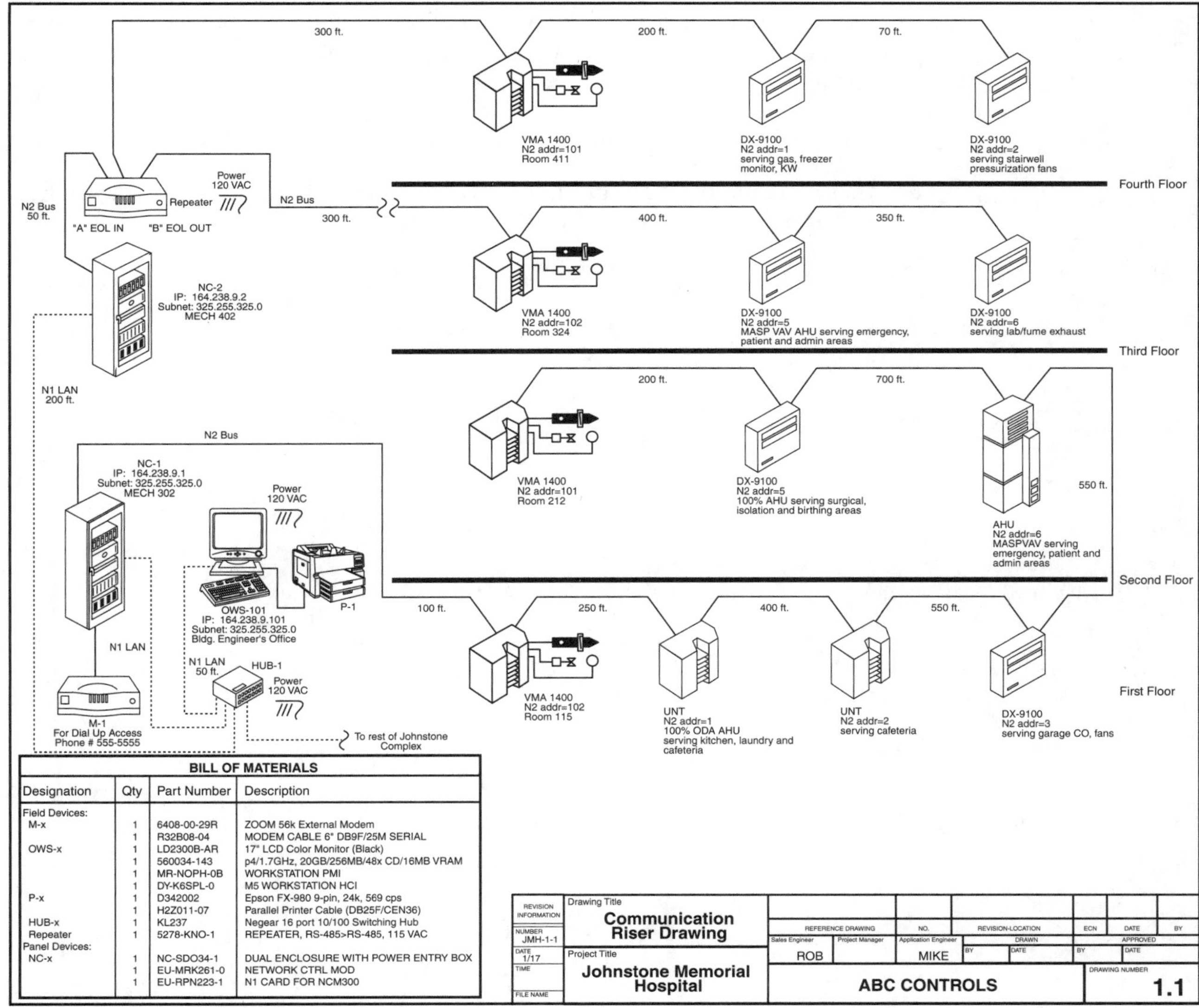

BILL OF MATERIALS

Designation	Qty	Part Number	Description
Field Devices:			
M-x	1	6408-00-29R	ZOOM 56k External Modem
	1	R32B08-04	MODEM CABLE 6" DB9F/25M SERIAL
OWS-x	1	LD2300B-AR	17" LCD Color Monitor (Black)
	1	560034-143	p4/1.7GHz, 20GB/256MB/48x CD/16MB VRAM
	1	MR-NOPH-0B	WORKSTATION PMI
	1	DY-K6SPL-0	M5 WORKSTATION HCI
P-x	1	D342002	Epson FX-980 9-pin, 24k, 569 cps
	1	H2Z011-07	Parallel Printer Cable (DB25F/CEN36)
HUB-x	1	KL237	Negear 16 port 10/100 Switching Hub
Repeater	1	5278-KNO-1	REPEATER, RS-485>RS-485, 115 VAC
Panel Devices:			
NC-x	1	NC-SDO34-1	DUAL ENCLOSURE WITH POWER ENTRY BOX
	1	EU-MRK261-0	NETWORK CTRL MOD
	1	EU-RPN223-1	N1 CARD FOR NCM300

_______________ **1.** Complete the command line to determine if the OWS-101 NIC card is operating properly. C:\>ping ___

_______________ **2.** Complete the command line to determine if NC-1 is communicating properly. C:\>ping ___

_______________ **3.** Complete the command line to determine if NC-2 is communicating properly. C:\>ping ___

_______________ **4.** The subnet mask for OWS-101 is ___.

_______________ **5.** The subnet mask for NC-1 is ___.

_______________ **6.** The subnet mask for NC-2 is ___.

_______________ **7.** Are the subnet masks for OWS-101, NC-1, and NC-2 the same?

_______________ **8.** Consideration has been given to expanding the building automation system by adding another workstation. A possible IP address for the new workstation OWS-102 is ___.

_______________ **9.** A possible subnet address for the new workstation OWS-102 is ___.

_______________ **10.** Consideration has been given to expanding the building automation system by adding another NC. A possible IP address for the new NC, NC-3 is ___.

_______________ **11.** A possible subnet address for the new NC, NC-3 is ___.

Section 22.1 Direct Digital Control Strategies

REVIEW QUESTIONS

Name ______________________________ **Date** ____________________

Multiple Choice

______________ **1.** ___ control is control in which feedback occurs between the controller, sensor, and controlled device.

A. Open loop
B. Closed loop
C. High/low signal
D. Universal input-output

______________ **2.** ___ is the difference between a control point and a setpoint.

A. Offset
B. Lead/lag
C. Algorithm
D. Tuning

______________ **3.** Setback is the unoccupied ___ setpoint.

A. cooling
B. heating
C. humidity
D. static pressure

______________ **4.** ___ control is a direct digital control feature that calculates an average value from all selected inputs.

A. Averaging
B. Proportional
C. Integral
D. Derivative

______________ **5.** ___ control algorithms are commonly used in basic control systems that do not require precise control.

A. Averaging
B. Proportional
C. Integral
D. Derivative

_______________ **6.** ___ is the downloading of the proper response times into a controller and checking the response of the control system.
A. Overshooting
B. Undershooting
C. Strategies
D. Tuning

_______________ **7.** The building automation system receives feedback on the control of the temperature in the building space by ___.
A. measuring the result of the action taken and changing the position or state of the controlled device
B. sending information sent through crossover cables back to the central processing unit
C. comparing the sensors in the building space to the sensors near the heating coil
D. checking for continuity in the closed loop system

_______________ **8.** The most common DDC feature is the ability to ___ in a building automation system.
A. lockout the economizer
B. maintain a setpoint
C. control the high- and low-limit setpoints
D. automatically reset the setpoints

_______________ **9.** ___ control is commonly used with mixed air damper controls.
A. Reset
B. Low-limit
C. High-limit
D. Lead/lag

_______________ **10.** An economizer is used at 68°F and below and is ___ above 68°F.
A. reversed
B. overworked
C. optimized
D. locked out

_______________ **11.** The most common application of ___ is the control of building space temperature at different locations in a building.
A. high/low signal select
B. averaging control
C. low-limit control
D. lead/lag control

_______________ **12.** Malfunction of one component in a closed loop control system results in ___.
A. an alarm signal
B. system failure
C. improper feedack between the controller and controlled device
D. other components within the system having the incorrect value or position

______________ **13.** The communication between a personal computer and building automation system controller CPU is referred to as ___.
A. data sharing
B. downloading the controller
C. feedback
D. networking

______________ **14.** Adaptive control algorithms reduce the amount of ___.
A. control loops in a BAS
B. time it takes for a controller to recieve feedback
C. time a technician tunes a control loop
D. maintenance checks and tune-ups

______________ **15.** ___ measure temperature, pressure, or humidity and provide the measured values as input to a controller.
A. Sensors
B. Relays
C. Tranducers
D. Open loops

______________ **16.** A(n) ___ is the desired value to be maintained by a system.
A. offset
B. setpoint
C. control point
D. setback

Completion

______________ **1.** A(n) ___ is a building automation system software method used to control the energy-using equipment in a building.

______________ **2.** ___ is the measurement of the results of a controller action by a sensor or switch.

______________ **3.** A(n) ___ is an HVAC system that uses outside air for cooling.

______________ **4.** ___ is a direct digital control feature in which a primary setpoint is reset automatically as another value (reset variable) changes.

______________ **5.** ___ is the alternation of operation between two or more similar pieces of equipment.

______________ **6.** ___ is a direct digital control feature in which a building automation system selects among the highest or lowest values from multiple inputs.

______________ **7.** An algorithm is a mathematical equation used by a building automation system controller to determine a desired ___.

______________ **8.** A(n) ___ control algorithm is a control algorithm that positions the controlled device in direct response to the amount of offset in a building automation system.

______________ **9.** In the HVAC industry, only extremely sensitive control applications require ___ control.

_______________ **10.** A(n) ___ control algorithm is a control algorithm that automatically adjusts its response time based on environmental conditions.

_______________ **11.** ___ control is control in which no feedback occurs between the controller, sensor, and controlled device.

_______________ **12.** ___ is the unoccupied cooling setpoint.

_______________ **13.** A(n) ___ is a description of the amount a reset variable resets the primary setpoint.

_______________ **14.** ___ features include setpoint control, reset control, low-limit control, high-limit control, economizer lockout control, lead/lag control, high/low signal select, and averaging control.

_______________ **15.** A direct digital control system is a control system in which the building automation system controller is wired directly to ___ and can turn them ON and OFF, or start a motor.

_______________ **16.** An integral control algorithm is a control algorithm that eliminates any ___ after a certain length of time.

_______________ **17.** When using the dry bulb economizer method, a(n) ___ is used to control economizer operation.

_______________ **18.** Setup/setback setpoint control saves energy by preventing heating and/or cooling systems from operating when a commercial building is ___.

_______________ **19.** A(n) ___ is the actual value that a control system experiences at any given time in a system.

Proportional Control

_______________ **1.** setpoint

_______________ **2.** offset

_______________ **3.** control point

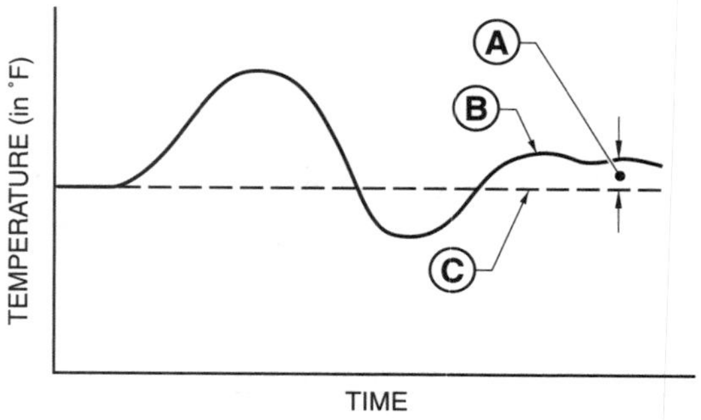

TEMPERATURE (in °F)
A
B
C
TIME

Section 22.2 Direct Digital Control System Applications

REVIEW QUESTIONS

Name ______________________________ **Date** ______________

Multiple Choice

______________ **1.** Common DDC system applications include ___ control of rooftop packaged units.
A. air handling unit
B. universal input-output
C. unitary
D. low-limit

______________ **2.** ___ control is used to prevent damage caused by freezing if the mixed air temperature becomes too low.
A. Low-limit
B. Reset
C. Lockout
D. Proportional

______________ **3.** The difference in volume is ___ related to the internal pressure in the building.
A. directly
B. indirectly
C. proportionally
D. inversely

______________ **4.** The universal input-output controller uses a(n) ___ to maintain a hot water temperature setpoint.
A. three-way hot water valve
B. electric/pneumatic tranducer
C. hot water supply temperature sensor
D. water supply thermometer

______________ **5.** The air handling unit controller provides a(n) ___ output signal to two variable-speed drives.
A. 2 mA to 10 mA
B. 4 mA to 20 mA
C. 8 mA to 40 mA
D. 10 mA to 50 mA

Completion

______________________ **1.** When using a rooftop packaged unit direct digital control system, a(n) ___ is used in place of a standard thermostat control.

______________________ **2.** In a hot water heating DDC system, a(n) ___ controller provides an output signal to an electric/pneumatic transducer.

______________________ **3.** In a variable-air-volume (VAV) DDC system, an air handling unit controller uses input from a(n) ___ to control supply and return fans through variable-speed drives.

Unitary Control of Rooftop Packaged Unit

_____________________ **1.** building space temperature sensor

_____________________ **2.** supply fan

_____________________ **3.** return air

_____________________ **4.** airflow switch

_____________________ **5.** compressor

_____________________ **6.** unitary controller

_____________________ **7.** heating element

_____________________ **8.** supply airflow

_____________________ **9.** filter

_____________________ **10.** airflow switch paddle

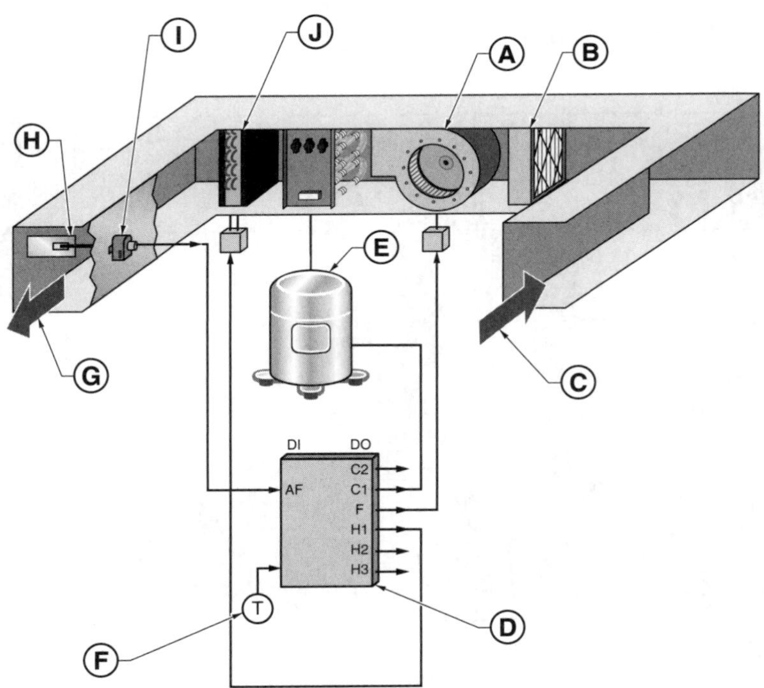

Air Handling Unit Control of Variable-Air-Volume System

______________ 1. normally open damper

______________ 2. static pressure sensor

______________ 3. return fan

______________ 4. airflow station with electronic transducer

______________ 5. variable-speed drive

______________ 6. supply fan

______________ 7. air handling unit controller

______________ 8. normally closed damper

______________ 9. outside air

______________ 10. return air

______________ 11. exhaust air

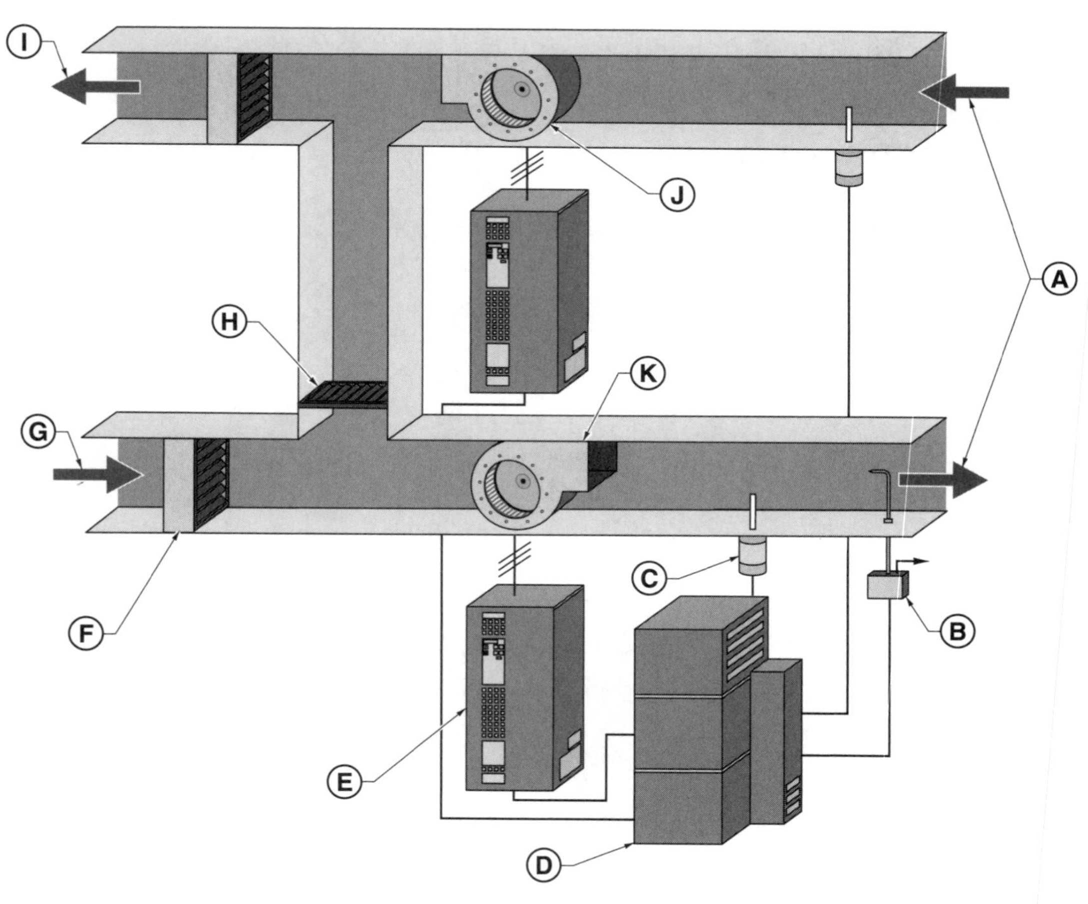

Name ______________________________ Date ____________________

Activity 22-1. Rooftop Air Handling Unit Control Strategies

Use the rooftop air handling unit sequence of operation to answer the questions.

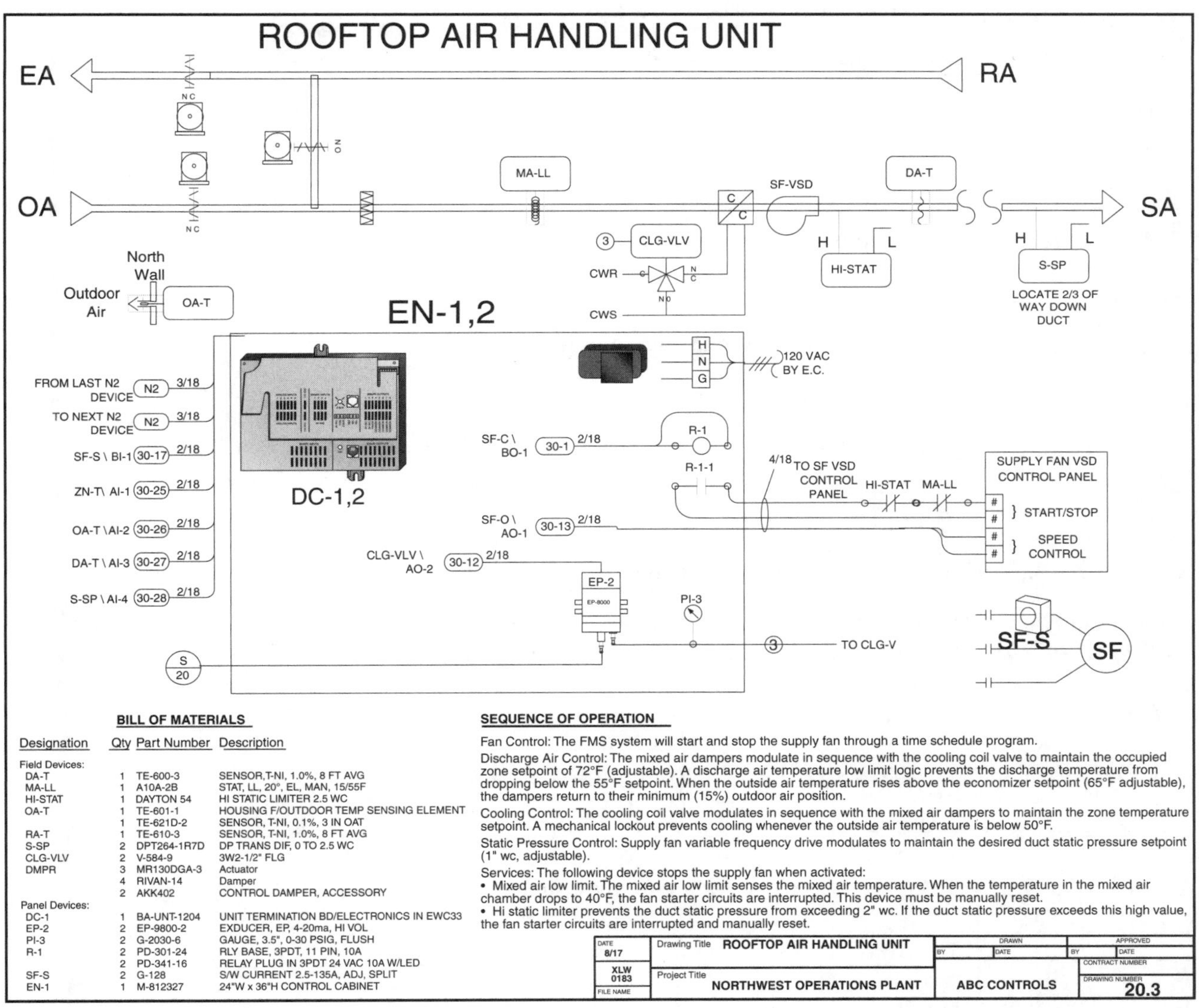

BILL OF MATERIALS

Designation	Qty	Part Number	Description
Field Devices:			
DA-T	1	TE-600-3	SENSOR,T-NI, 1.0%, 8 FT AVG
MA-LL	1	A10A-2B	STAT, LL, 20°, EL, MAN, 15/55F
HI-STAT	1	DAYTON 54	HI STATIC LIMITER 2.5 WC
OA-T	1	TE-601-1	HOUSING F/OUTDOOR TEMP SENSING ELEMENT
	1	TE-621D-2	SENSOR, T-NI, 0.1%, 3 IN OAT
RA-T	1	TE-610-3	SENSOR, T-NI, 1.0%, 8 FT AVG
S-SP	2	DPT264-1R7D	DP TRANS DIF, 0 TO 2.5 WC
CLG-VLV	2	V-584-9	3W2-1/2" FLG
DMPR	3	MR130DGA-3	Actuator
	4	RIVAN-14	Damper
	2	AKK402	CONTROL DAMPER, ACCESSORY
Panel Devices:			
DC-1	1	BA-UNT-1204	UNIT TERMINATION BD/ELECTRONICS IN EWC33
EP-2	2	EP-9800-2	EXDUCER, EP, 4-20ma, HI VOL
PI-3	2	G-2030-6	GAUGE, 3.5", 0-30 PSIG, FLUSH
R-1	2	PD-301-24	RLY BASE, 3PDT, 11 PIN, 10A
	2	PD-341-16	RELAY PLUG IN 3PDT 24 VAC 10A W/LED
SF-S	2	G-128	S/W CURRENT 2.5-135A, ADJ, SPLIT
EN-1	1	M-812327	24"W x 36"H CONTROL CABINET

SEQUENCE OF OPERATION

Fan Control: The FMS system will start and stop the supply fan through a time schedule program.

Discharge Air Control: The mixed air dampers modulate in sequence with the cooling coil valve to maintain the occupied zone setpoint of 72°F (adjustable). A discharge air temperature low limit logic prevents the discharge temperature from dropping below the 55°F setpoint. When the outside air temperature rises above the economizer setpoint (65°F adjustable), the dampers return to their minimum (15%) outdoor air position.

Cooling Control: The cooling coil valve modulates in sequence with the mixed air dampers to maintain the zone temperature setpoint. A mechanical lockout prevents cooling whenever the outside air temperature is below 50°F.

Static Pressure Control: Supply fan variable frequency drive modulates to maintain the desired duct static pressure setpoint (1" wc, adjustable).

Services: The following device stops the supply fan when activated:
- Mixed air low limit. The mixed air low limit senses the mixed air temperature. When the temperature in the mixed air chamber drops to 40°F, the fan starter circuits are interrupted. This device must be manually reset.
- Hi static limiter prevents the duct static pressure from exceeding 2" wc. If the duct static pressure exceeds this high value, the fan starter circuits are interrupted and manually reset.

DATE 8/17	Drawing Title ROOFTOP AIR HANDLING UNIT	DRAWN BY / DATE	APPROVED BY / DATE
XLW 0183 / FILE NAME	Project Title NORTHWEST OPERATIONS PLANT	ABC CONTROLS	CONTRACT NUMBER / DRAWING NUMBER 20.3

______________ **1.** Is the starting and stopping of the supply fan based on a DDC strategy?

______________ **2.** If it is not based on temperature, what is it based on?

_______________ **3.** The occupied zone setpoint is ___°F.

_______________ **4.** Is the occupied zone setpoint adjustable?

5. What devices are modulated in sequence to maintain the cooling setpoint?

____

_______________ **6.** What DDC strategy prevents the discharge air temperature from dropping too low?

_______________ **7.** The discharge air temperature setpoint is ___°F.

_______________ **8.** What DDC strategy determines if outside air can provide cooling?

_______________ **9.** What type of DDC strategy is it?

_______________ **10.** The outside air setpoint is ___°F.

11. How can it be determined that humidity control is not involved?

____

____

12. What DDC strategy maintains static pressure in the supply duct?

____

_______________ **13.** The static pressure setpoint is ___.

_______________ **14.** Is the static pressure setpoint adjustable?

_______________ **15.** Is the mixed air low limit connected to the rooftop controller?

_______________ **16.** Is any averaging or reset control indicated?

Activity 22-2. Rooftop Air Handling Unit Service Call

A too-hot complaint is received from the area supplied by the rooftop air handling unit. In addition, the occupants claim that the unit is often noisy. After investigation, it is discovered that the duct static pressure is hunting (cycling) and not maintaining setpoint. Duct noise increases as the static pressure increases. A laptop computer or service tool is connected to the controller.

1. List the components of the closed loop control for the fan static pressure.

____

After examination, it is determined that the static loop parameters were downloaded but never checked.

_______________ **2.** What might the static pressure control loop need?

3. List the basic steps of how this might be done.

____

____

Activity 22-3. Hot Water Boiler Control Strategies

Use the hot water boiler control sequence of operation to answer the questions.

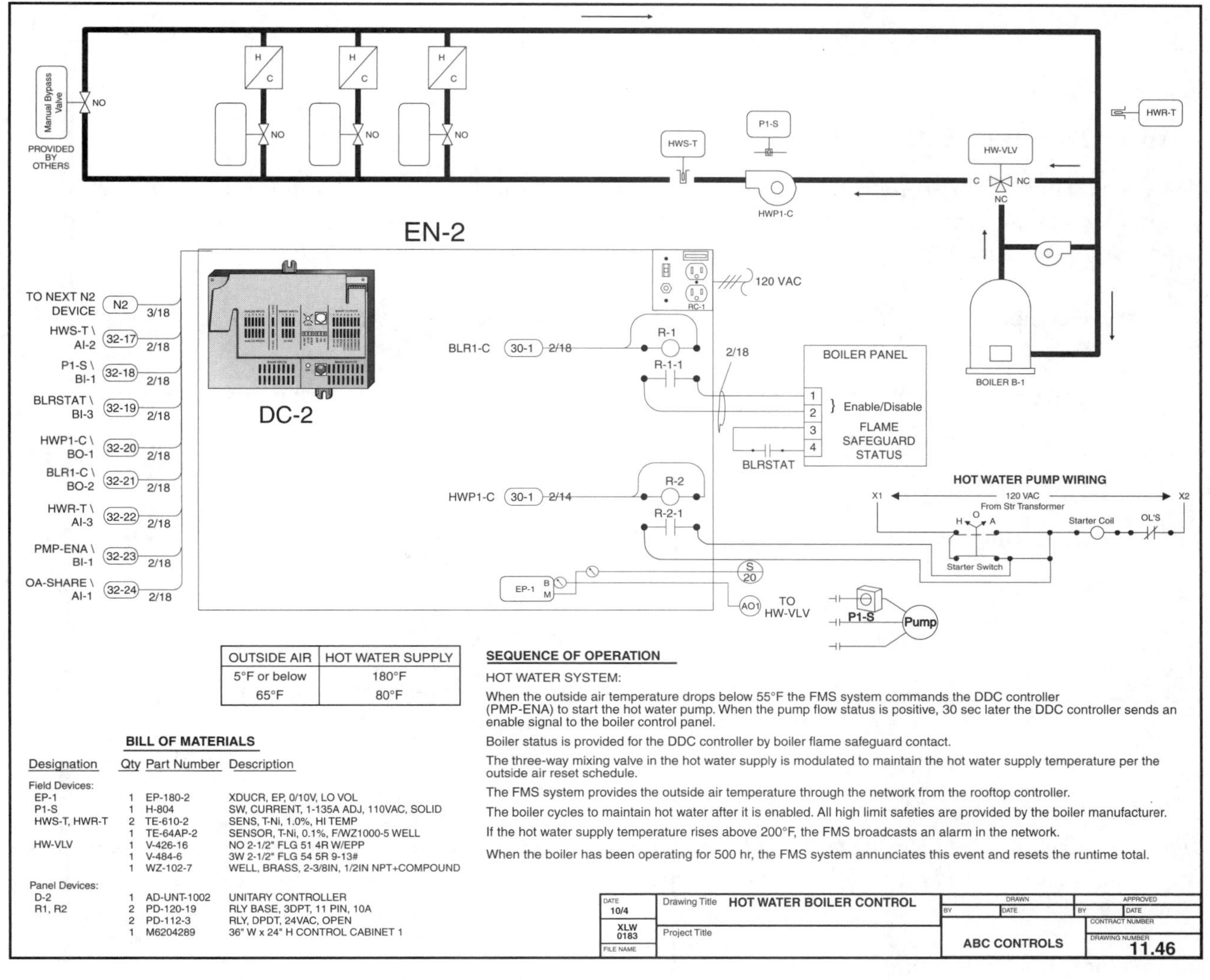

____________________ **1.** What starts and stops pump HWP1-C?

____________________ **2.** What is its setpoint?

____________________ **3.** What device provides boiler status?

____________________ **4.** What device is modulated to maintain the correct outside air temperature in the system?

____________________ **5.** Does the hot water supply temperature have a fixed setpoint?

____________________ **6.** What DDC strategy changes the hot water setpoint as the outside air temperature changes?

____________________ **7.** The reset schedule outside air temperature values are ___°F and ___°F.

________________ **8.** The reset schedule hot water supply temperature values are ___°F and ___°F.

________________ **9.** Where does the outside air temperature value come from?

________________ **10.** Is there an alarm setpoint?

________________ **11.** If so, the alarm setpoint value is ___°F.

Activity 22-4. Hot Water Boiler Service Call

A too-cold complaint is received from the area supplied by the hot water boiler. After investigation, it is determined that the area is too cold. The hot water pump and boiler are running, and the hot water supply temperature is warm. A laptop computer or service tool is connected to the controller.

1. List the components of the closed loop control for the hot water supply temperature.

____

____

The outside air temperature is checked and determined to be 35°F. The hot water supply setpoint is given as 150°F.

________________ **2.** Is the hot water supply setpoint correct?

________________ **3.** Using the reset schedule on the hot water boiler control print, an approximate hot water setpoint at the given outside air temperature is ___°F.

Upon checking the reset schedule in the software, it is discovered that the wrong values were entered. After correcting the values, it is noticed that the hot water three-way valve is constantly opening and closing.

4. List three things that may cause this condition.

____

____

Upon further checking, it is discovered that the integration value is 0.

5. Should this value be changed? Why or why not?

____

____

Section 23.1 Life-Safety and Time-Based Supervisory Control

REVIEW QUESTIONS

Name ______________________________ Date ________________

Multiple Choice

______________ **1.** ___ supervisory control is a control strategy for life safety issues such as fire prevention, detection, and suppression.

A. Time-based
B. Optimum start/stop
C. Duty cycling
D. Life safety

______________ **2.** ___ provides the ability for building automation systems to join loads that are used during the same time.

A. Duty cycling
B. Schedule linking
C. Temporary scheduling
D. Alternate scheduling

______________ **3.** Time-based start time represents the ___.

A. actual start time of a unit
B. amount of time it takes to bring a space to the desired temperature
C. beginning of building occupancy
D. programmed start time of a unit

______________ **4.** ___ enables building loads to be scheduled for use for more than one unique time schedule per year.

A. Alternate scheduling
B. Schedule linking
C. Holiday and vacation scheduling
D. Time zone scheduling

______________ **5.** A(n) ___ is a time-based supervisory control strategy in which the occupants can change a zone from an unoccupied to occupied mode for temporary occupancy.

A. seven-day program
B. temporary schedule
C. alternate schedule
D. timed override

_______________ **6.** Most building automation system software provides ___ separate programmable time periods per day.
A. two or three
B. four
C. a maximum of five
D. up to six

_______________ **7.** A 5+2 time-based supervisory control strategy recognizes ___.
A. holidays that fall on a normal workday (Monday through Friday)
B. separate start times for two different employee shifts
C. Monday through Friday as normal workdays and Saturday and Sunday as additional workdays
D. Monday through Friday as normal workdays and Saturday and Sunday separately

_______________ **8.** ___ was created to reduce the workload of HVAC technicians.
A. Life-safety supervisory control
B. Time-based supervisory control
C. Control strategy priority
D. Shift scheduling

Completion

_______________ **1.** A(n) ___ is a programmable software method used to control the energy-consuming functions of a commercial building.

_______________ **2.** ___ allows an HVAC technician to individually program building automation system ON and OFF time functions for each day of the week.

_______________ **3.** A(n) ___ is a holiday that changes its date each year.

_______________ **4.** ___ supervisory control is a control strategy in which the time of day or day of the week is used to determine the desired operation of a load.

_______________ **5.** ___ is required when multiple strategies are combined for an HVAC unit.

_______________ **6.** Supervisory control strategies and ___ are commonly used for the same energy devices during the same period.

_______________ **7.** ___ is a supervisory control strategy in which the HVAC load is turned on as late as possible to obtain the proper building space temperature at the beginning of building occupancy.

REVIEW QUESTIONS

Name ______________________________ Date ____________________

Multiple Choice

______________ **1.** ___ compensates for a specific event in a building without using a timed override.
A. Schedule linking
B. Saving time changeover
C. Alternate scheduling
D. Temporary scheduling

______________ **2.** ___ values can be used as indicators of HVAC equipment efficiency and/or mechanical problems.
A. Estimation control
B. Duty cycling
C. Life safety supervisory
D. Thermal recovery coefficient

______________ **3.** ___ is the percentage of time a load or circuit is ON compared to the time a load or circuit is ON and OFF (total cycle time).
A. Duty cycle
B. Estimation control
C. Timed override
D. Saving time changeover

______________ **4.** ___ control is a control method that adjusts (learns) its control settings based on the condition of a building.
A. Adaptive start
B. Estimation
C. Optimum start
D. Duty cycle

______________ **5.** A(n) ___ is the ratio of an indoor temperature change and the length of time it takes to obtain that temperature change.
A. optimized factor
B. duty cycle
C. optimum start/stop strategy
D. thermal recovery coefficient

_______________ **6.** The optimum stop supervisory control strategy turns an HVAC unit OFF ___.
- A. once optimum building temperature has been reached
- B. before the end of building occupancy
- C. after the end of building occupancy
- D. once the thermal recovery coefficient has been reached

Completion

_______________ **1.** ___ is a supervisory control strategy in which the HVAC load is turned ON as late as possible to obtain the proper building space temperature at the beginning of building occupancy.

_______________ **2.** ___ is a control method that adjusts (learns) its control settings based on the condition of a building.

_______________ **3.** The ___ supervisory control strategy is designed to reduce electrical demand in a commercial building.

_______________ **4.** ___ is a control method that uses the latest building temperature data to estimate the actual start time to heat or cool a building before occupancy.

Duty Cycling

_______________ **1.** load 2 ON

_______________ **2.** one load ON at any given time

_______________ **3.** load 1 ON

_______________ **4.** load 2 OFF

_______________ **5.** load 1 OFF

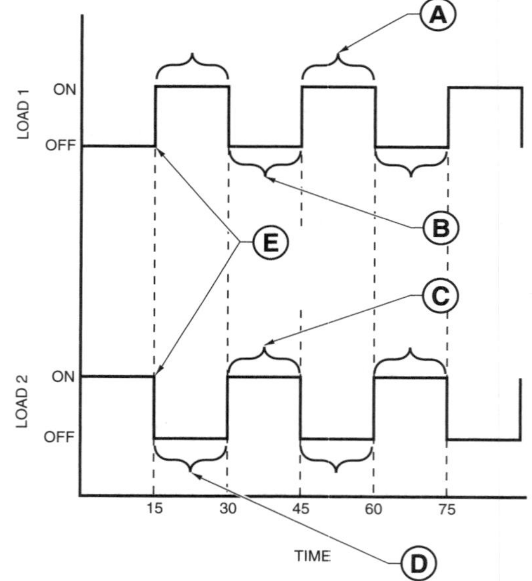

MULTIPLE LOADS

Section 23.3 Electrical-Demand Supervisory Control

REVIEW QUESTIONS

Name ______________________________ **Date** ______________________

Multiple Choice

_______________ **1.** A ___ load is an electric load that has been turned OFF by an electrical demand supervisory control strategy.
A. low-priority
B. targeted
C. restored
D. shed

_______________ **2.** A ___ load is a shed load that has been turned ON by an electrical demand supervisory control strategy.
A. low-priority
B. targeted
C. restored
D. changeover

_______________ **3.** A ___ load is a load that is shed first when a high electrical demand period occurs.
A. restored
B. minimum
C. low-priority
D. high-priority

_______________ **4.** In ___ shedding, load 1 is the first load shed the first time a high electrical demand condition occurs and load 2 is the first load shed the next time a high electrical demand condition occurs.
A. fixed load
B. rotating priority load
C. primary load table
D. effective load

_______________ **5.** The ___ time timer causes a load to be restored after it has been shed for a certain length of time.
A. maximum shed
B. minimum shed
C. maximum load
D. minimum load

______________ 6. An effective electrical-demand supervisory control strategy requires accurate ___ electrical demand targets.
   - A. weekly
   - B. monthly
   - C. quarterly
   - D. annual

Completion

______________ 1. ___ is the highest amount of electricity used during a specific period of time.

______________ 2. A(n) ___ is a table that prioritizes the order in which electrical loads are turned OFF.

______________ 3. A(n) ___ load is a load that is important to the operation of a building and is shed last when demand goes up.

______________ 4. ___ is an electrical demand supervisory control strategy in which the order of loads to be shed is changed with each high electrical demand condition.

______________ 5. When building electrical demand is above the target, the loads in the ___ shed table are shed in order.

Electrical Demand Control

______________ 1. shortest length of time load will operate

______________ 2. minimum length of time load is OFF

______________ 3. maximum length of time load is OFF

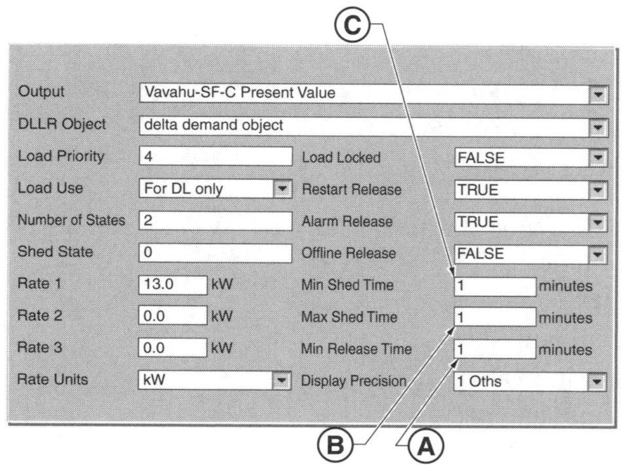

Name ______________________________ Date ______________

Activity 23-1. Rooftop Air Handling Unit Control Strategies

Use the rooftop air handling unit sequence of operation to answer the questions.

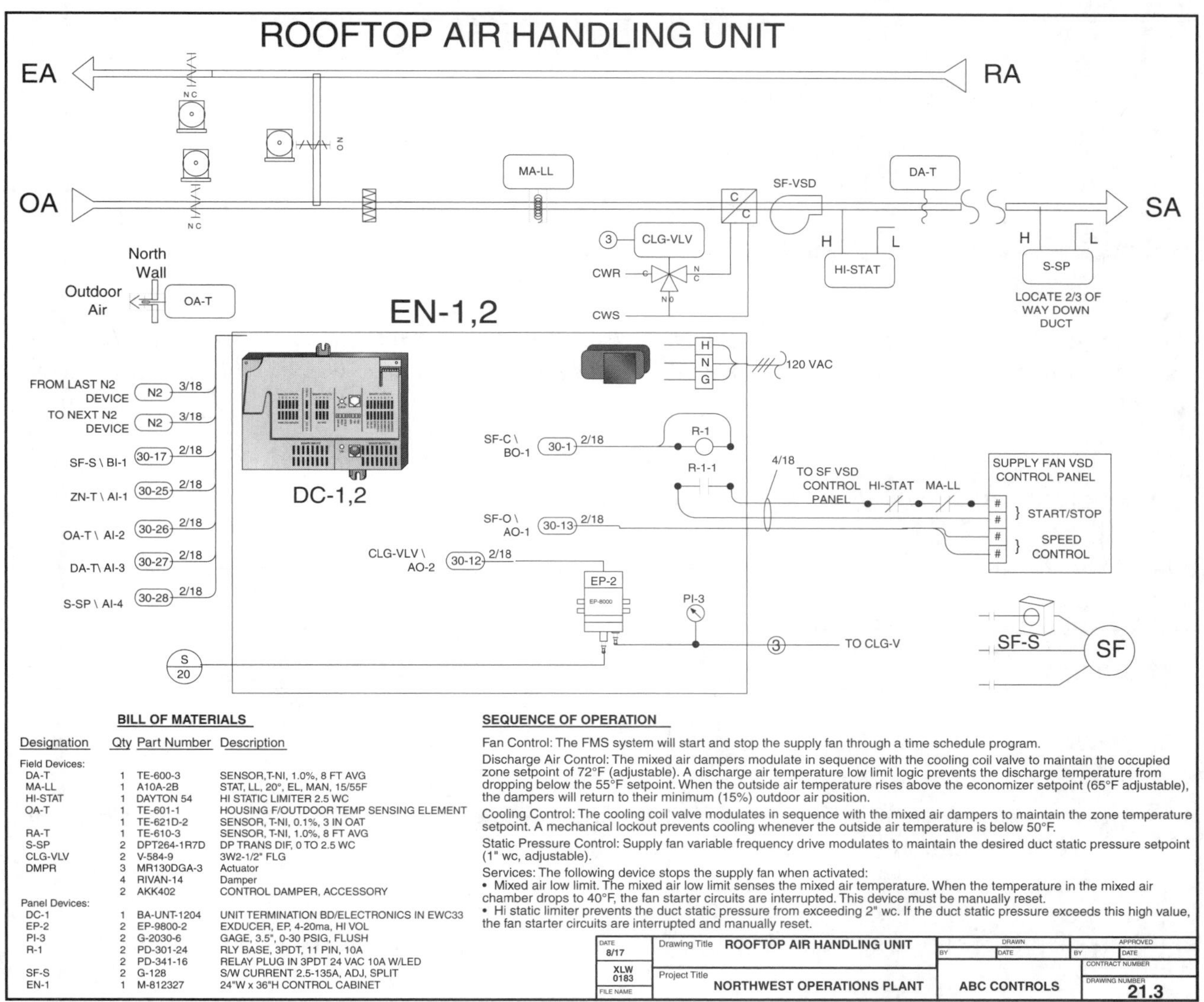

_______________ **1.** Is the starting and stopping of the supply fan based on a supervisory control strategy?

_______________ **2.** What supervisory control strategy is the starting and stopping of the supply fan based on?

_______________ **3.** If the air handling unit were in a hospital, would this control strategy be used?

_______________ **4.** What device located at the air handling unit uses electrical power?

5. What factors may be taken into consideration when considering demand control for the unit?

____

____

6. If the air handling unit static pressure setpoint were reduced, what would happen to the electrical demand at the supply fan?

____

Activity 23-2. Rooftop Air Handling Unit Service Call

On Wednesday at 11 AM, a too-hot complaint is received from the area supplied by the rooftop air handling unit. Upon arrival, it is determined that the unit is off. A laptop computer or service tool is connected to the controller.

1. What supervisory control strategies may be keeping the unit off?

____

____

____

After checking the software, it is determined that the time schedule was incorrectly set to turn the unit off at 10 AM instead of 10 PM. After changing the value, the unit starts.

2. What other time scheduling features may be checked to ensure that they are not incorrectly set?

____

____

____

Activity 23-3. Single Duct VAV Box Control Strategies

Use the VAV box sequence of operation to answer the questions.

VAV BOX SEQUENCE OF OPERATION

UNOCCUPIED MODE:

THE VAV BOX IS PUT INTO UNOCCUPIED MODE BY A TIME SCHEDULE IN THE BUILDING DDC SYSTEM. IN THIS MODE, THE VAV BOX MAINTAINS THE BUILDING SPACE CONDITIONS TO THE SETUP/SETBACK TEMPERATURE SETPOINTS. THE BUILDING SPACE TEMPERATURE SENSOR MODULATES THE BOX DAMPER DEPENDING ON BUILDING SPACE REQUIREMENTS.

OCCUPIED MODE:

THE VAV BOX IS PUT INTO OCCUPIED MODE BY A TIME SCHEDULE IN THE BUILDING DDC SYSTEM. IN THIS MODE, THE VAV BOX MODULATES THE VARIABLE VOLUME DAMPER TO MAINTAIN THE BUILDING SPACE CONDITIONS AT THE OCCUPIED TEMPERATURE SETPOINTS.

ON A CALL FOR COOLING, THE VARIABLE VOLUME DAMPER IS MODULATED FROM MINIMUM AIRFLOW TO MAXIMUM COOLING AIR FLOW TO MAINTAIN BUILDING SPACE CONDITIONS. ON A DROP IN BUILDING SPACE TEMPERATURE, THE VARIABLE VOLUME DAMPER IS MODULATED TO ITS MINIMUM FLOW POSITION. ON A FURTHER CALL FOR HEAT, THE ELECTRIC HEAT CYCLES ON TO MAINTAIN SETPOINT.

SERIES FAN:

THE SERIES FAN IS OFF DURING THE SHUTDOWN MODE. THE FAN IS ALWAYS ON DURING THE OCCUPIED MODE AND IS CYCLED ON DURING THE UNOCCUPIED MODE.

THE ON/OFF SERIES FAN IS CONTROLLED BY A SINGLE BINARY OUTPUT WITH MINIMUM ON/OFF TIMERS THAT ARE ADJUSTABLE.

RADIANT HEATING CONTROL:

THE PERIMETER AREAS HAVE FINNED-TUBE RADIANT HEATING. THE CONTROL OF THE FINNED-TUBE HEATERS COMES FROM THE PERIMETER VAV CONTROLLERS. TYPICALLY, ON A CALL FOR HEAT THE DAMPER OF THE VAV CLOSES AND RECIRCULATES PRECONDITIONED BUILDING SPACE AIR. ON A CONTINUED CALL FOR HEAT, THE TWO-POSITION VALVE ON THE RADIANT FINNED TUBES OPENS AS THE BUILDING SPACE TEMPERATURE BECOMES SATISFIED, THE VALVES CLOSE.

______________ **1.** What supervisory strategy puts the VAV box into unoccupied mode?

2. What happens to the temperature setpoints?

____

____

3. What does the VAV box fan do during the occupied mode?

____

4. What does the VAV box fan do differently during the unoccupied mode?

____

____

______________ **5.** If the VAV box controller has a minimum airflow setpoint of 200 cfm during the occupied mode, would the same amount of air be needed in the unoccupied mode?

______________ **6.** Does the system include electric heating elements?

Activity 23-4. Single Duct VAV Box Service Call

A too-hot complaint is received from the area supplied by a single duct VAV box. After investigation, it is determined that the room temperature is 85°F. A laptop computer or service tool is connected to the controller. The laptop or service tool indicates that the occupied cooling setpoint is 75°F, the unoccupied cooling setpoint is 85°F, and the current setpoint is 85°F.

______________________ **1.** What mode is the VAV box in?

______________________ **2.** What should be changed?

______________________ **3.** Where should it be changed?

Section 24.1 Resources for Programming

REVIEW QUESTIONS

Name ______________________ **Date** ______________

Multiple Choice

_______________ **1.** The ___ may be stored in a central control area or in an electrical cabinet at the controller.

A. on-site laptop
B. RS-232 cables or RJ-45 cables
C. job prints
D. printer drivers

_______________ **2.** ___ uses a wireless interface box that plugs into a jack on the field controller or wall sensor.

A. Bluetooth®
B. RJ-45 cables (Ethernet 8 pin cables)
C. Embedded programing
D. Graphical user interface (GUI) programming

_______________ **3.** The ___ will often indicate the desired operation of a controller in different modes, such as economizer, heating, cooling, air flow, or pressure setpoints.

A. embedded programming
B. commissioning agent
C. written sequence of operation
D. sensor package

_______________ **4.** Since a technician is expected to support both old as well as brand new BASs, the service laptops may have a variety of ___.

A. operating systems
B. interface cables
C. user interfaces
D. troubleshooting documents

_______________ **5.** For controllers that are difficult to access, it is preferable to ___.

A. connect an interface box to the field controller
B. use Ethernet 8-pin cables
C. pass a field controller through the web-enabled supervisory controller
D. use a Bluetooth connection

______________ **6.** There is a(n) ___ technology cycle in the BAS industry in which hardware, software, and new devices are quickly introduced.
A. 3 to 9 month
B. 6 to 12 month
C. 12 to 18 month
D. 18 to 24 month

Completion

______________ **1.** The ___ in the job prints will often indicate the desired operation of a controller in different modes.

______________ **2.** The ___ should show controller wiring and interfacing to electrical devices such as motors, compressors, and transducers.

______________ **3.** It is preferable to use the ___ method for controllers that are difficult to access.

______________ **4.** When using ___, problems can arise because the technician is not physically at the controller and cannot directly observe the controller or mechanical equipment.

______________ **5.** A(n) ___ may be expected to load and maintain controller programming software tools on a customer's computer.

______________ **6.** If the ___ is vague or does not cover needed details of the programming, a technician will have to "fill in the blanks" for the desired controller programming.

Section 24.2 Programming Methods

REVIEW QUESTIONS

Name ______________________________ **Date** ____________________

Multiple Choice

______________ **1.** An advantage of text-based programming is that it offers ___.

A. program recovery in the event of controller damage
B. quick debugging of the programming
C. simple syntax and sequences of operation
D. a large amount of flexibility in custom programming

______________ **2.** In embedded programming, the setpoints may be changed through the ___.

A. BAS or at a local keypad menu
B. controller bus
C. wireless Bluetooth connection
D. remote keypad menu or controller

______________ **3.** When a controller is programmed as a(n) ___, the controller will be used to connect inputs and outputs without much control logic.

A. embedded program device
B. communication bus
C. remote controller
D. generic multiplexer

______________ **4.** Examples of a(n) ___ include temperature sensors for freezers and coolers as well as multiple exhaust fans and lighting loads.

A. list with radio buttons
B. generic multiplexer
C. supervisory controller
D. embedded chip controller

______________ **5.** Modification is the most difficult with ___ programming.

A. fully programmable GUI
B. menu-driven
C. embedded
D. text-based controller

______________ **6.** With ___ programming, a programming tool is used to build a controller program.

A. text-based
B. menu-driven
C. fully programmable GUI
D. embedded

__________ **7.** A disadvantage of menu-driven programming is that ___.
- A. a program may need to be built for each field controller
- B. the programming language is difficult to learn
- C. troubleshooting can be complicated
- D. modification is difficult

__________ **8.** In fully programmable graphical-user-interface programming, ___ are connected together to form functions and sequences of operation such as start/stop loops.
- A. multiplexers
- B. coding sequences
- C. field controllers
- D. graphical blocks

Completion

__________ **1.** ___ is a text language that uses a number of control statements such as "if-then" and "go to."

__________ **2.** With a GUI programming tool, controller functions are added as ___ in the programming.

__________ **3.** ___ an existing program means that the basic controller program will not be changed but that a feature will be changed or added.

__________ **4.** A disadvantage of ___ programming is that there are many syntax rules that must be followed exactly.

__________ **5.** ___ programming is normally reserved for complex or customized applications.

__________ **6.** In today's BAS industry, the three main programming types are embedded, ___, and fully programmable GUI programming.

Section 24.3 Commissioning and Troubleshooting

REVIEW QUESTIONS

Name ______________________________ **Date** ____________________

Multiple Choice

______________ **1.** In ___ mode, the software can be checked on the computer only without actually running the HVAC equipment.
A. troubleshooting
B. simulation
C. commissioning
D. test

______________ **2.** Common troubleshooting issues include mode, setpoint, and ___.
A. simulation failure
B. failed switching from heating to cooling
C. modified programming
D. debugging

______________ **3.** After a service technician takes action to rectify a customer complaint, the technician must ___ to ensure that the system is working properly.
A. run a simulation
B. review the modified programming
C. restart the system
D. show the customer the cause of the issue

______________ **4.** Ideally, a VAV air handler runs at ___″ wc.
A. ½
B. ¾
C. 1
D. 1½

______________ **5.** In general, the capabilities of the fully programmable GUI programming are only limited by the ___.
A. complexity of the application
B. mechanical equipment available
C. skill and experience of the programmer
D. inflexibility in nonstandard applications

______________ **6.** Troubleshooting requires resources such as a ___.
A. Bluetooth connection and a time schedule
B. reheat valve and an economizer
C. copy of the controller program and a desktop computer
D. laptop and control drawings

______________ **7.** The simulation mode may include a feature that will allow the program to ___.
A. complete the written sequences without manual intervention
B. run at a faster speed
C. operate at a slower speed to allow the technician to inspect each part of the written sequences
D. check the software while the HVAC equipment is running

______________ **8.** In a VAV terminal box, the minimum CFM setting is usually ___ of the maximum.
A. 10% to 30%
B. 30% to 50%
C. 50% to 70%
D. 70% to 90%

Completion

______________ **1.** Troubleshooting involves receiving a service call, uploading the controller, and then ___ the program.

______________ **2.** Controller ___ is performed to confirm that a controller program works as it is supposed to.

______________ **3.** It may be desirable to ___ the equipment outside normal building occupancy hours if possible.

ACTIVITIES

Name ______________________________ Date ____________________

Fan-Powered VAV Box with Reheat

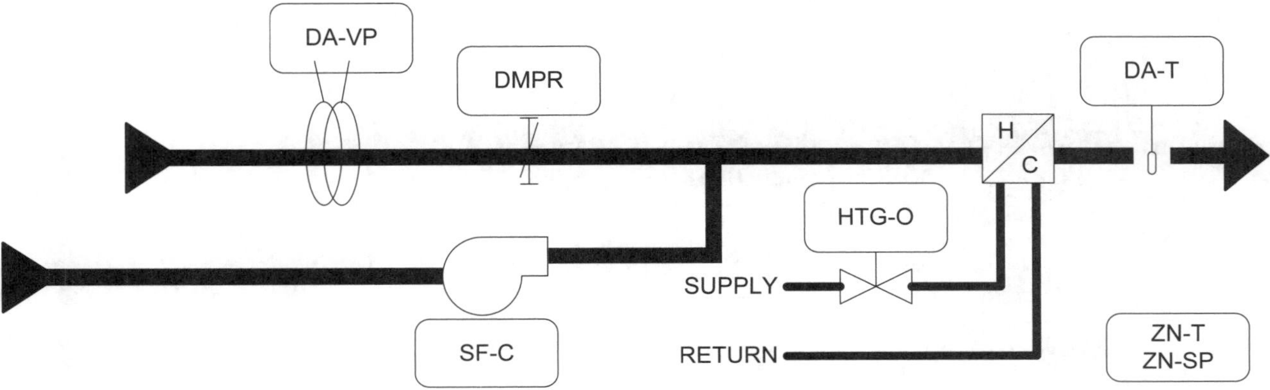

Operation. A DDC controller operates on a local communication bus with other controllers. It controls a pressure-independent variable-air-volume (VAV) box. The VAV box controls the temperature in a zone by varying the flow of air into the zone. On a rise in zone temperature above the cooling setpoint, the damper increases the airflow rate to a maximum of 500 CFM and the reheat valve modulates open. On a drop in zone temperature below the heating setpoint, the fan cycles ON while the reheat valve modulates open. During this heating cycle the damper is at the minimum airflow rate of 100 CFM. A discharge air temperature sensor is provided.

Occupied Mode. While occupied, the heating and cooling setpoints are 70°F and 74°F, respectively. The occupants can adjust the setpoints up or down by 3°F.

Unoccupied Mode. While unoccupied, the heating and cooling setpoints are 62°F and 82°F, respectively.

Activity 24-1. Basic Box Programming Information

Use the provided illustration to answer the programming questions.

1. Does this VAV box have hot water or electric reheat?

2. What type of heating actuator does it use?

3. Is a setpoint adjustment indicated for the occupants?

4. What type of supply damper is indicated?

5. Does this VAV box have a supply or discharge sensor?

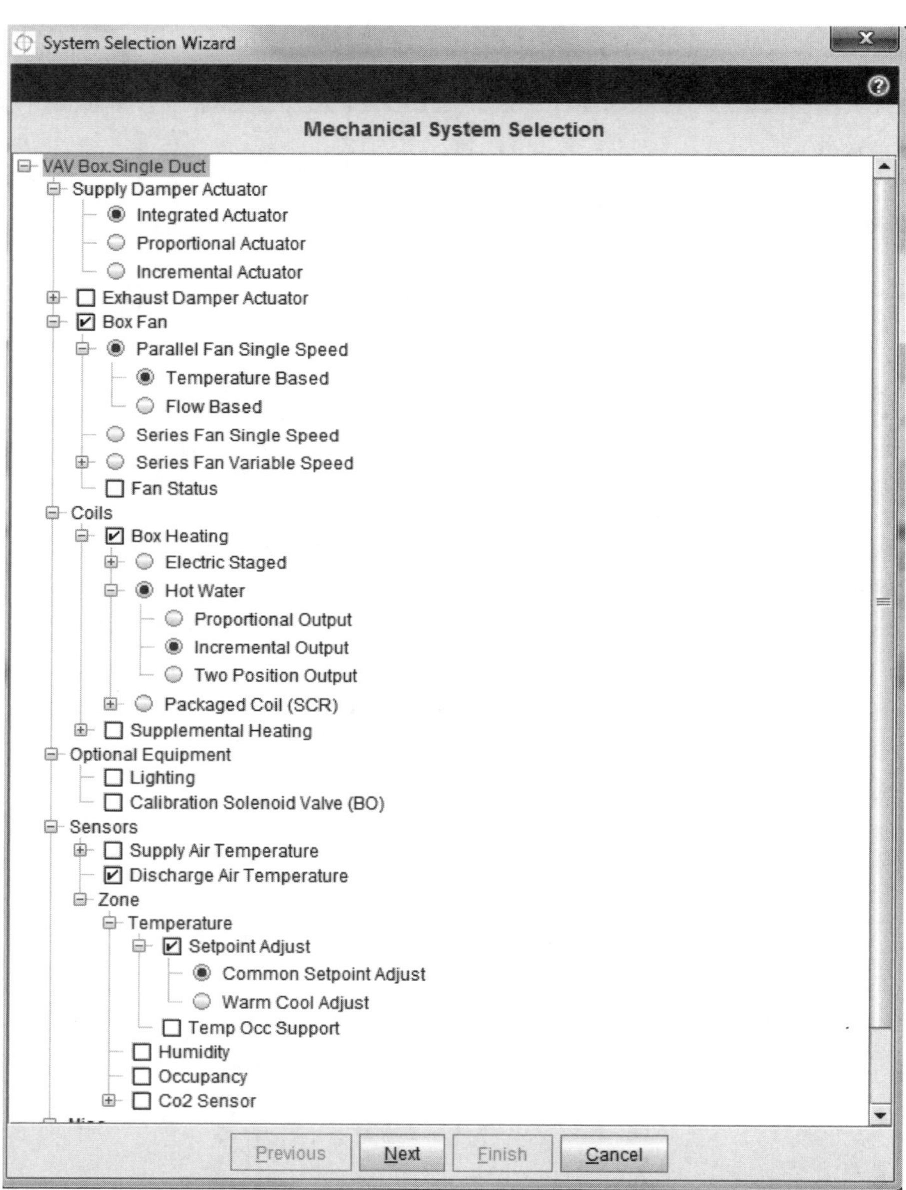

Activity 24-2. Temperature Setpoint Information

Use the provided illustration to answer the questions.

1. What value should be put in the CLG-OCC-SP?
2. What value should be put in the CLG-UNOCC-SP?
3. What value should be put in the HTG-OCC-SP?
4. What value should be put in the HTG-UNOCC-SP?

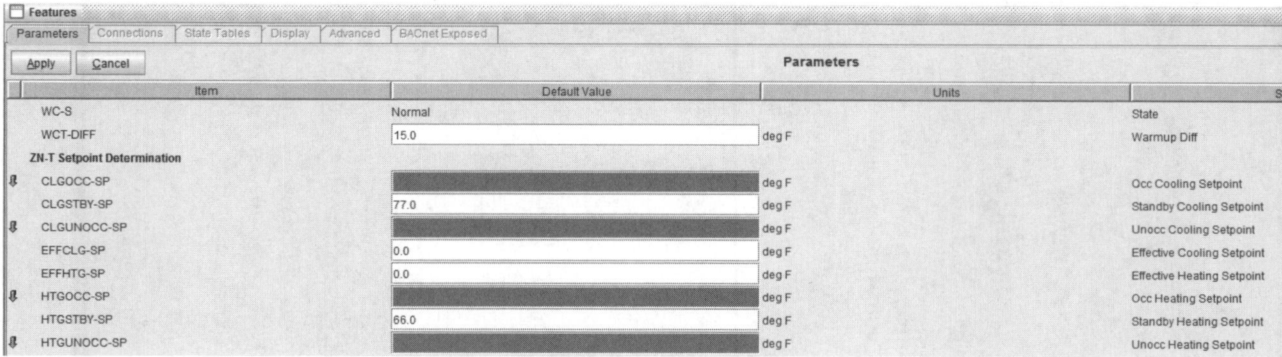

Activity 24-3. CFM Setpoint Information

Use the provided illustration to answer the questions.

1. What value should be put in the CLG-MAXFLOW?
2. What value should be put in the CLGOCC-MINFLOW?
3. What value should be put in the HTG-OCCMINFLOW?
4. What value is indicated for the supply area of this VAV box?

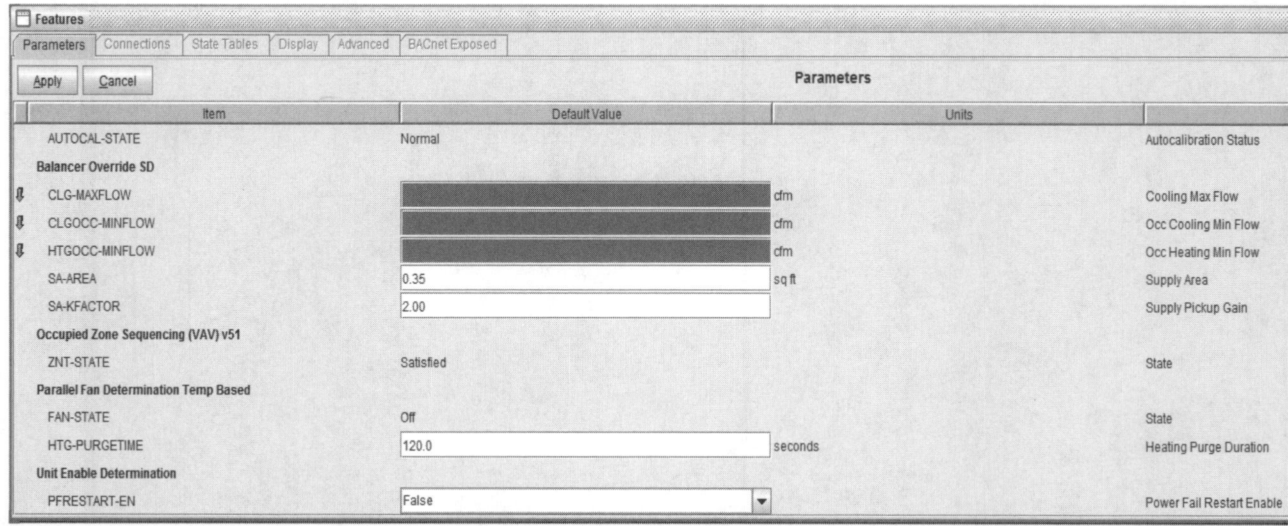

Section 25.1 Building Automation Retrofit

REVIEW QUESTIONS

Name ______________________________ **Date** ______________________

Multiple Choice

______________ **1.** The ___ for a rooftop packaged unit retrofit application includes a zone temperature sensor package with override, a power supply package, the rooftop unit wiring diagram, and a riser diagram for controller addressing.
A. pneumatic controller
B. building survey
C. database
D. materials list

______________ **2.** Many hot water boiler systems have a ___ which controls the temperature of the water pumped through the building heating coils.
A. pressure valve
B. three-way valve
C. 120 V control circuit
D. digital interface

______________ **3.** Most new chillers have a built-in ___.
A. cooling tower
B. water circulation system
C. microprocessor control panel
D. power supply system

______________ **4.** A common alternate retrofit strategy for a chiller control system is to ___.
A. use variable frequency drives in place of vane volume control
B. reset the fan static setpoint based on zone demand
C. install a flame safeguard controller that is compatible with the BAS communication bus
D. install a translator panel to allow the BAS to view and command points inside the microprocessor control panel

______________ **5.** A boiler control retrofit application normally consists of retrofitting an existing ___ boiler control system to a building automation system.
A. electromechanical
B. ON/OFF
C. digital
D. three-phase

______________ **6.** A(n) ___ is the process of upgrading building controls and mechanical systems with a building automation system.

A. modification
B. rebuild
C. retrofit
D. upgrade

______________ **7.** A(n) ___ is an inventory of the energy-consuming equipment in a commercial building.

A. building survey
B. materials list
C. energy survey
D. equipment list

Completion

______________ **1.** A(n) ___ terminal box controller is a controller that modulates the damper inside a VAV terminal box to maintain a specific building space temperature.

______________ **2.** A(n) ___ combines heating and cooling operations in one piece of equipment for most commercial buildings.

______________ **3.** A(n) ___ air handling unit is an air handling unit that moves a variable volume of air.

______________ **4.** When retrofitting a new building automation system to an older existing system, it is common to use a(n) ___ that allows the new building automation system to view and adjust the old building system.

______________ **5.** A(n) ___ is burner control equipment that monitors a burner start-up sequence and the main flame during normal operation, and provides an air purge to rid the combustion chamber of unburned fuel during a shutdown.

______________ **6.** A(n) ___ may be either wired as an analog output from the BAS or commanded directly from the BAS if it is compatible with the system manufacturer.

______________ **7.** ___ is a more realistic measurement of outside air for ventilation and total airflow than velocity pressure.

______________ **8.** A(n) ___ or software bridge may enable points in the microprocessor control panel to be viewed, alarmed, or even controlled by the BAS.

______________ **9.** If multiple boilers are present, they should be put in sequence to avoid improper ___.

Section 25.2 Retrofit Applications

REVIEW QUESTIONS

Name ______________________________ Date ____________________

Multiple Choice

_______________ **1.** A(n) ___ is the process of upgrading a building automation system by replacing obsolete or worn parts, components, and controls with new or modern ones.
- A. commissioning
- B. retrofit
- C. download
- D. interface setup

_______________ **2.** ___ is the process of adjusting and balancing an HVAC system for maximum efficiency.
- A. Commissioning
- B. Point mapping
- C. Interface setup
- D. System links

_______________ **3.** The ___ of retrofitting are the building survey, documentation, and planning; existing control removal; new building automation system installation; programming and commissioning; human-computer interface setup; and ongoing support.
- A. system troubleshooting steps
- B. processes
- C. stages
- D. systems

Completion

_______________ **1.** A(n) ___ is an inventory of the energy-consuming equipment in a commercial building.

_______________ **2.** ___ indicates the sequences of operation, lists of control points and locations, setpoints, strategies, and interlocks for the building automation system.

_______________ **3.** A(n) ___ is the completed programming information of a controller.

_______________ **4.** A(n) ___ is the central computer that enables the building staff to view the operation of the building automation system.

_______________ **5.** ___ is the process of adding the individual input and output points to the database of the human-computer interface.

_______________ **6.** ___ is the process of sending a controller database from a personal computer to a controller.

______________________ **7.** ___ are used for many retrofit tasks, so efficient project management, planning, and coordination are required for a successful installation.

______________________ **8.** A(n) ___ is a device that receives a signal from a sensor, compares it to a setpoint value, and sends an appropriate output signal to a controlled device.

______________________ **9.** ___ is the creation of software.

______________________ **10.** Components containing ___ such as mercury must be disposed of properly, following all environmental regulations.

______________________ **11.** A(n) ___ includes the specific controllers, sensors, power supplies, cabling, and other hardware required for the new building automation system installation.

______________________ **12.** Components that are ___ in a retrofit are covered under the original warranty.

Rooftop Packaged Unit Controllers

______________________ **1.** electronic controller

______________________ **2.** power fuse block

______________________ **3.** temperature sensor termination block

______________________ **4.** control system transformer

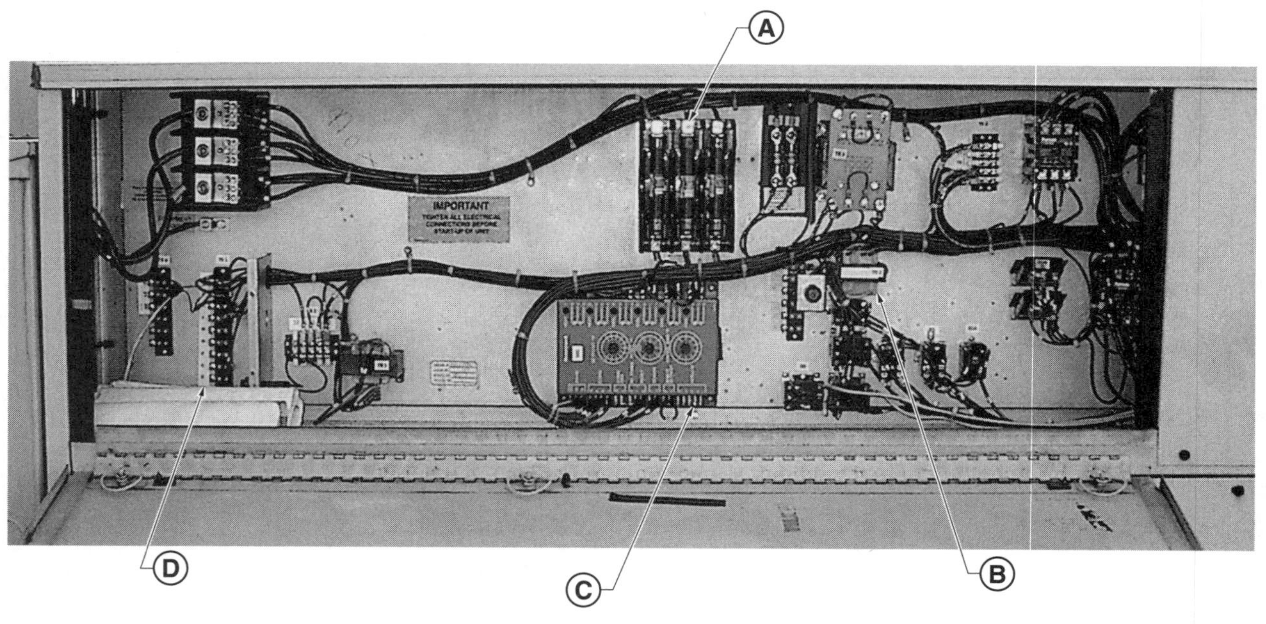

Building Automation System Retrofit of Existing Systems

ACTIVITIES

Name ______________________________ **Date** ______________________

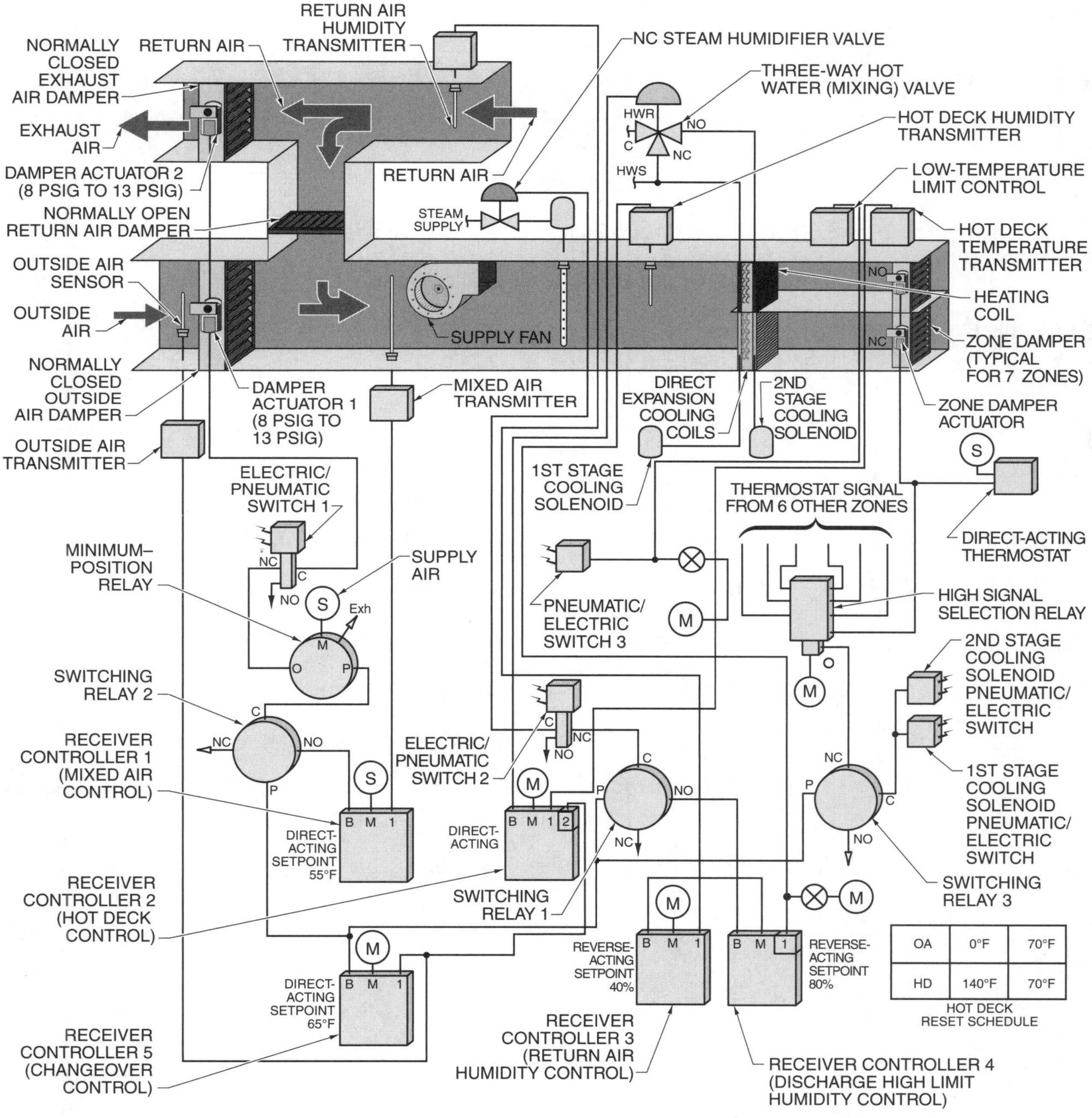

RETURN AIR HUMIDITY TRANSMITTER
NORMALLY CLOSED EXHAUST AIR DAMPER
RETURN AIR
NC STEAM HUMIDIFIER VALVE
THREE-WAY HOT WATER (MIXING) VALVE
HWR
NO
C
NC
HWS
HOT DECK HUMIDITY TRANSMITTER
EXHAUST AIR
DAMPER ACTUATOR 2 (8 PSIG TO 13 PSIG)
RETURN AIR
LOW-TEMPERATURE LIMIT CONTROL
NORMALLY OPEN RETURN AIR DAMPER
STEAM SUPPLY
HOT DECK TEMPERATURE TRANSMITTER
OUTSIDE AIR SENSOR
NO
HEATING COIL
OUTSIDE AIR
SUPPLY FAN
NC
ZONE DAMPER (TYPICAL FOR 7 ZONES)
NORMALLY CLOSED OUTSIDE AIR DAMPER
DAMPER ACTUATOR 1 (8 PSIG TO 13 PSIG)
MIXED AIR TRANSMITTER
DIRECT EXPANSION COOLING COILS
2ND STAGE COOLING SOLENOID
ZONE DAMPER ACTUATOR
OUTSIDE AIR TRANSMITTER
S
ELECTRIC/ PNEUMATIC SWITCH 1
1ST STAGE COOLING SOLENOID
THERMOSTAT SIGNAL FROM 6 OTHER ZONES
DIRECT-ACTING THERMOSTAT
MINIMUM–POSITION RELAY
NC
C
NO
SUPPLY AIR
S
Exh
M
O
P
PNEUMATIC/ ELECTRIC SWITCH 3
M
HIGH SIGNAL SELECTION RELAY
O
M
SWITCHING RELAY 2
C
NC
NO
P
2ND STAGE COOLING SOLENOID PNEUMATIC/ ELECTRIC SWITCH
RECEIVER CONTROLLER 1 (MIXED AIR CONTROL)
S
ELECTRIC/ PNEUMATIC SWITCH 2
C
NC
NO
M
C
P
NO
NC
NC
P
C
NO
1ST STAGE COOLING SOLENOID PNEUMATIC/ ELECTRIC SWITCH
B M 1
DIRECT-ACTING SETPOINT 55°F
DIRECT-ACTING
B M 1 2
RECEIVER CONTROLLER 2 (HOT DECK CONTROL)
SWITCHING RELAY 1
SWITCHING RELAY 3
M
M
M
M
DIRECT-ACTING SETPOINT 65°F
B M 1
REVERSE-ACTING SETPOINT 40%
B M 1
B M 1
REVERSE-ACTING SETPOINT 80%

OA	0°F	70°F
HD	140°F	70°F

HOT DECK RESET SCHEDULE

RECEIVER CONTROLLER 5 (CHANGEOVER CONTROL)
RECEIVER CONTROLLER 3 (RETURN AIR HUMIDITY CONTROL)
RECEIVER CONTROLLER 4 (DISCHARGE HIGH LIMIT HUMIDITY CONTROL)

Activity 25-1. Mechanical System Identification

A building contains a number of pneumatically-controlled air handling units that must be retrofitted to direct digital control. Use the multizone air handling unit drawing to answer the questions.

______________ **1.** The air handling unit type is ___.

______________ **2.** The air handling unit supplies conditioned air to ___ (number) zone(s).

______________ **3.** The air handling unit contains ___ (number) fan(s).

4. List the air handling unit pneumatic actuators.

____

____

______________ **5.** The total number of actuators in the system is ___.

______________ **6.** The number of pneumatic temperature sensors is ___.

7. List the binary (ON/OFF) devices in the system.

____

____

______________ **8.** The control strategy needed for the hot deck is ___.

______________ **9.** Two control strategies that may be used with the outside air dampers are ___.

______________ **10.** Two control strategies that may be used with the humidifier are ___.

______________ **11.** The control strategy used to control the mechanical cooling is ___.

______________ **12.** The number of electric/pneumatic transducers needed is ___.

______________ **13.** At a cost of $55 per transducer, the total transducer cost is $___.

______________ **14.** If 1 hr of installation time is needed per transducer, at $75/hr, the installation cost of the electric/pneumatic transducers is $___.

______________ **15.** If each temperature sensor costs $80, the total temperature sensor cost is $___.

Section 26.1 Building System Management

REVIEW QUESTIONS

Name ______________________________ **Date** ______________

Multiple Choice

______ **1.** The most common alarms used in commercial buildings are associated with ___ sensors.
A. temperature
B. humidity
C. static pressure
D. flows

______ **2.** ___ is the use of past building equipment performance data to determine future system needs.
A. Commissioning
B. Data trending
C. Preventive maintenance
D. Troubleshooting

______ **3.** A time interval of ___ min is commonly used for long-term data trending.
A. 10
B. 20
C. 30
D. 60

______ **4.** A ___ is created to record building conditions and make a decision regarding the purchase or replacement of HVAC units.
A. maintenance file
B. data log
C. cost analysis
D. data trend

______ **5.** Short time intervals such as ___ can indicate improper equipment operation and are used to troubleshoot equipment problems.
A. 1 min or 2 min
B. 2 min or 3 min
C. 3 min or 4 min
D. 4 min or 5 min

______________ 6. Personal computers that perform building system management functions are known as ___.
- A. off-site monitoring devices
- B. human computer interfaces
- C. preventive maintenance hardware
- D. personal maintenance monitors

Completion

______________ 1. A(n) ___ function provides information to maintenance technicians who maintain or manage a commercial building.

______________ 2. ___ is the most common building system management function used in commercial buildings.

______________ 3. A(n) ___ is the amount of change required in a variable for the alarm to return to normal after it has been in alarm status.

______________ 4. The most common operator interface device that receives alarm notification is a(n) ___.

______________ 5. ___ can be set up to monitor most inputs and outputs of a building automation system.

______________ 6. In ___, values and conditions of a commercial building are recorded during a specific time interval.

______________ 7. ___ alarms concern elements of building operation that are not vital.

______________ 8. ___ can be imported into a spreadsheet program and used to create graphs and charts.

______________ 9. Alarms are used to alert a maintenance technician of ___.

______________ 10. ___ alarms concern devices vital to proper operation of a commercial building.

______________ 11. The ___ feature gives an HVAC unit time to operate before a change in alarm status.

______________ 12. Data trends can be used to change the ___ of a unit or correct mechanical equipment problems.

Section 26.2 Building Maintenance

REVIEW QUESTIONS

Name ______________________________ Date ____________________

Multiple Choice

______________ **1.** ___ maintenance is the monitoring of wear conditions and equipment characteristics and comparing them to a predetermined tolerance to predict possible malfunctions or failures.
- A. Routine
- B. Computer
- C. Preventive
- D. Predictive

______________ **2.** All building automation systems are capable of producing ___.
- A. an integrated view of the building system
- B. a list of the cost for each device in the system
- C. alarms in the event of device malfunction
- D. documentation of the information provided to technicians

______________ **3.** ___ maintenance is performed after equipment has failed.
- A. Predictive
- B. Preventive
- C. Corrective
- D. Routine

______________ **4.** A ___ has the ability to add hourly labor cost factors and determine the preventive maintenance dollar value performed by each technician.
- A. building documentation system
- B. graphic software program
- C. computerized maintenance management system
- D. data trending report

______________ **5.** ___ maintenance attempts to detect equipment problems before failure occur by use of vibration and other sensors.
- A. Predictive
- B. Preventive
- C. Corrective
- D. Routine

Completion

______________ **1.** ___ maintenance is scheduled inspection and work (lubrication, adjustment, cleaning) required to maintain equipment in peak operating condition.

______________ **2.** Preventive maintenance is usually less expensive than the ___ approach.

______________ **3.** Building automation systems use graphics software which ___ communicates building and equipment conditions to maintenance technicians.

______________ **4.** Some building management systems use ___ of a particular piece of equipment so that actual temperature, humidity, pressure, or equipment status values can be superimposed.

______________ **5.** The building system management documentation function is commonly used to record ___.

______________ **6.** Preventive maintenance software can create preventive maintenance work orders associated with specific ___.

Preventive Maintenance Software

______________ **1.** maintenance information

______________ **2.** equipment identification

______________ **3.** expected variable range (possible alarm setpoints)

______________ **4.** monitored operating variable

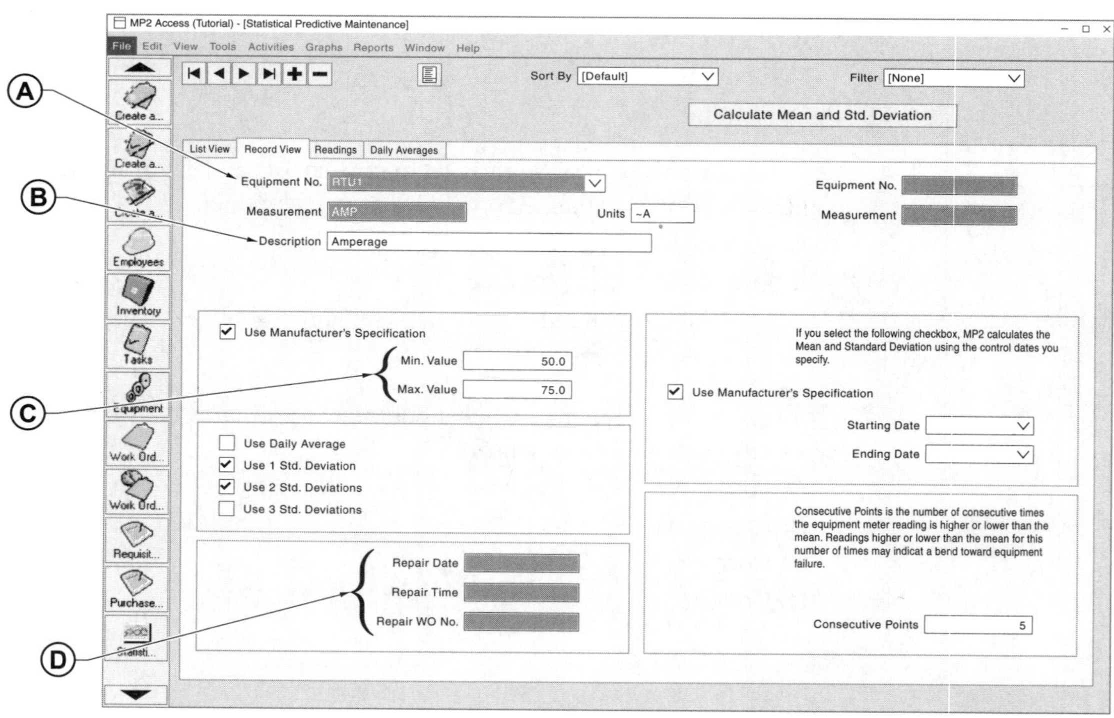

ACTIVITIES

Name ______________________________ Date ____________________

Activity 26-1. Alarm Value Setup

Use the rooftop air handling unit drawing to answer the questions.

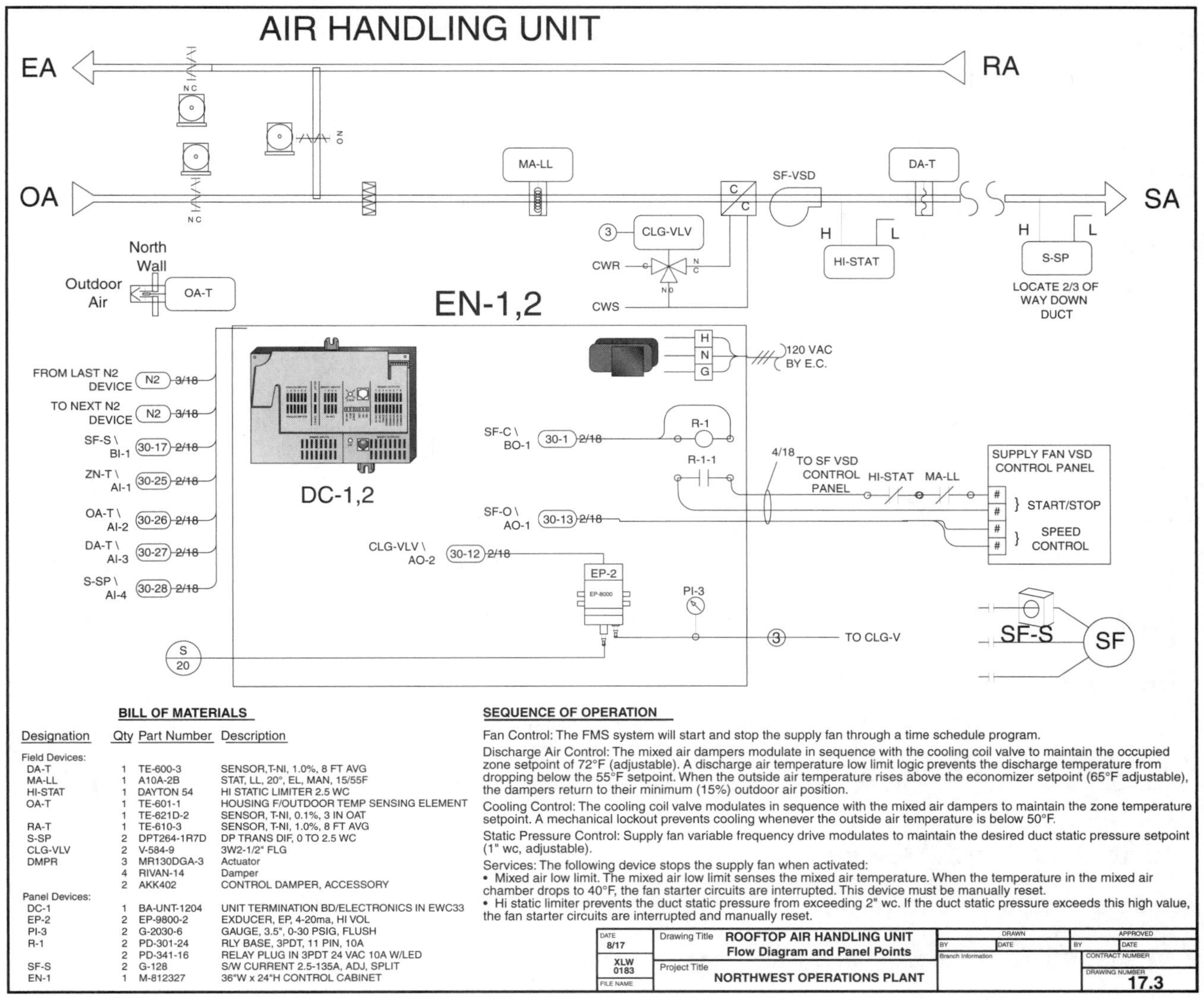

BILL OF MATERIALS

Designation	Qty	Part Number	Description
Field Devices:			
DA-T	1	TE-600-3	SENSOR,T-NI, 1.0%, 8 FT AVG
MA-LL	1	A10A-2B	STAT, LL, 20°, EL, MAN, 15/55F
HI-STAT	1	DAYTON 54	HI STATIC LIMITER 2.5 WC
OA-T	1	TE-601-1	HOUSING F/OUTDOOR TEMP SENSING ELEMENT
	1	TE-621D-2	SENSOR, T-NI, 0.1%, 3 IN OAT
RA-T	1	TE-610-3	SENSOR, T-NI, 1.0%, 8 FT AVG
S-SP	2	DPT264-1R7D	DP TRANS DIF, 0 TO 2.5 WC
CLG-VLV	2	V-584-9	3W2-1/2" FLG
DMPR	3	MR130DGA-3	Actuator
	4	RIVAN-14	Damper
	2	AKK402	CONTROL DAMPER, ACCESSORY
Panel Devices:			
DC-1	1	BA-UNT-1204	UNIT TERMINATION BD/ELECTRONICS IN EWC33
EP-2	2	EP-9800-2	EXDUCER, EP, 4-20ma, HI VOL
PI-3	2	G-2030-6	GAUGE, 3.5", 0-30 PSIG, FLUSH
R-1	2	PD-301-24	RLY BASE, 3PDT, 11 PIN, 10A
	2	PD-341-16	RELAY PLUG IN 3PDT 24 VAC 10A W/LED
SF-S	2	G-128	S/W CURRENT 2.5-135A, ADJ, SPLIT
EN-1	1	M-812327	36"W x 24"H CONTROL CABINET

SEQUENCE OF OPERATION

Fan Control: The FMS system will start and stop the supply fan through a time schedule program.

Discharge Air Control: The mixed air dampers modulate in sequence with the cooling coil valve to maintain the occupied zone setpoint of 72°F (adjustable). A discharge air temperature low limit logic prevents the discharge temperature from dropping below the 55°F setpoint. When the outside air temperature rises above the economizer setpoint (65°F adjustable), the dampers return to their minimum (15%) outdoor air position.

Cooling Control: The cooling coil valve modulates in sequence with the mixed air dampers to maintain the zone temperature setpoint. A mechanical lockout prevents cooling whenever the outside air temperature is below 50°F.

Static Pressure Control: Supply fan variable frequency drive modulates to maintain the desired duct static pressure setpoint (1" wc, adjustable).

Services: The following device stops the supply fan when activated:
- Mixed air low limit. The mixed air low limit senses the mixed air temperature. When the temperature in the mixed air chamber drops to 40°F, the fan starter circuits are interrupted. This device must be manually reset.
- Hi static limiter prevents the duct static pressure from exceeding 2" wc. If the duct static pressure exceeds this high value, the fan starter circuits are interrupted and manually reset.

DATE 8/17	Drawing Title **ROOFTOP AIR HANDLING UNIT Flow Diagram and Panel Points**	DRAWN BY / DATE	APPROVED BY / DATE
XLW 0183 FILE NAME	Project Title **NORTHWEST OPERATIONS PLANT**	Branch Information	CONTRACT NUMBER / DRAWING NUMBER 17.3

1. What are the analog inputs for the air handling unit?

____

____

______________ **2.** Is a zone temperature occupied setpoint given in the sequence of operation?

______________ **3.** The zone temperature occupied setpoint is ___°F.

______________ **4.** The part number for the DA-T sensor is ___.

______________ **5.** The description given for EP-2 is ___.

______________ **6.** The length of the RA-T sensing element is ___ ft.

______________ **7.** There are two wires of ___ gauge used for the S-SP sensor.

______________ **8.** Is a duct static pressure setpoint given in the sequence of operation?

______________ **9.** The duct static pressure setpoint is ___″ wc.

______________ **10.** The mixed air low limit setpoint is ___°F.

______________ **11.** Which two devices can stop the supply fan?

______________ **12.** The outside air cooling lockout setpoint is ___°F.

______________ **13.** The range of the differential pressure transmitter is ___″ wc.

______________ **14.** Is a discharge air temperature setpoint given in the sequence of operation?

______________ **15.** The discharge air temperature setpoint is ___°F.

______________ **16.** The minimum position setting for the outside air damper is ___%.

______________ **17.** What will start and stop the supply fan?

______________ **18.** How many wires and what are the wire sizes for the N2 communication bus?

______________ **19.** What is the location of the S-SP (supply static pressure sensor)?

______________ **20.** What type of point is the fan status switch?

______________ **21.** Would an alarm be created for it?

______________ **22.** What position would the switch be in when in alarm?

Activity 26-2. Alarm Message Setup

Use the critical alarm notification to answer the questions.

1. Write an appropriate alarm message for the zone temperature.

____

____

2. Write an appropriate alarm message for the duct static pressure.

____

____

3. Write an appropriate alarm message for the discharge air temperature.

____

____

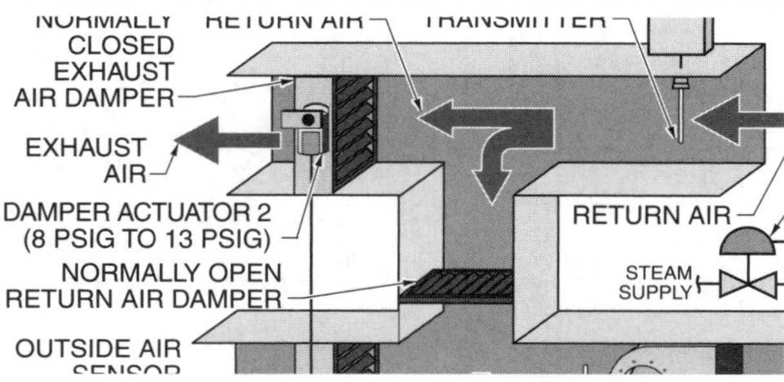

Activity 26-3. Using Data Trending

Use the air handling unit temperature data trends to answer the questions.

____________________ **1.** On what date did the lowest air-handling-unit outside air temperature occur?

____________________ **2.** What is the highest air-handling-unit zone temperature?

____________________ **3.** On which floor were all these temperature measurements taken?

____________________ **4.** Is the status of the supply fan indicated on this trend?

5. List at least three additional pieces of air-handling-unit information that may need to be trended.

____

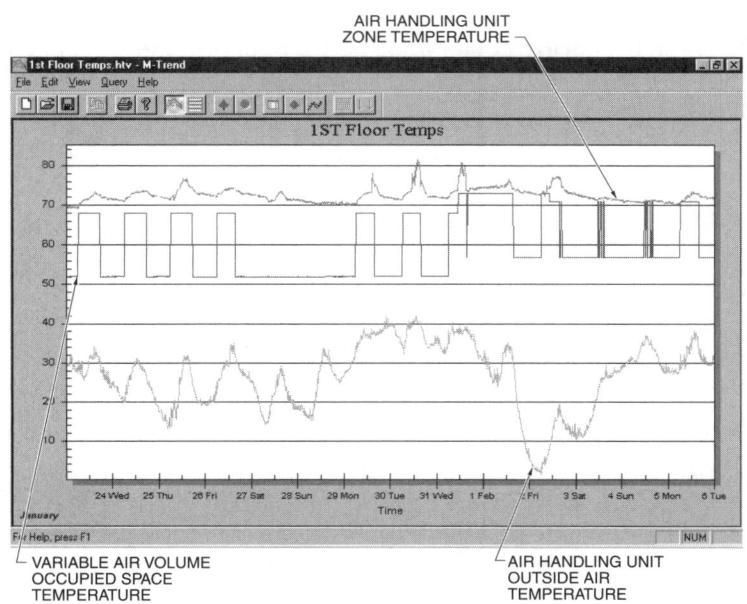

Activity 26-4. Preventive Maintenance Work Order Setup

Use the preventive maintenance work order and checklist to answer the questions.

PREVENTIVE MAINTENANCE WORK ORDER		
Work Order No: **46**	Requisition No:	
Issue Date: **6\21** Time: **15:49**	Skill: **Mechanical**	
Equipment Name: **Air Unit #3**	EQUIPMENT LOCATION	
Manufacturer: **Trane**	Building: **3E**	Floor: **3rd**
Description: **Air Handling Unit, 4 HP Century Motor**	System:	
Model No: **V4517** Serial No: **100-AHU01**	Date Work Performed:	

LUBRICATION	COMMENTS	
Grease motor bearings		☐
Grease fan bearings		☐
Oil damper pivots		☐
Oil damper pneumatics		☐

PREVENTIVE MAINTENANCE CHECKLIST		
MECHANICAL	COMMENTS	
Lockout/Tagout power supply		☐
Inspect motor rotation (bearings)		☐
Inspect fan rotation (bearings)		☐
Inspect fan blades		☐
Inspect V-belts		☐
Inspect motor and fan sheaves		☐
Inspect dampers		☐
Inspect damper actuators		☐
Replace unit filters		☐
Vacuum debris from unit		☐

1. List the tools required to perform the indicated preventive maintenance on the air handling unit.

____

____

2. List items consumed during preventive maintenance on an air handling unit.

____

____

3. List a minimum of three safety considerations when performing preventive maintenance on an air handling unit.

____

____

Section 27.1 Energy Auditing

REVIEW QUESTIONS

Name ____________________ **Date** ____________________

Multiple Choice

__________ **1.** ___ is the length of time it takes to recover the cost of an energy-saving measure.
- A. Return on investment
- B. Payback period
- C. Energy cost index
- D. Cost differential

__________ **2.** A ___ is a set of regulations that are related to a particular trade or environment.
- A. code
- B. standard
- C. rule
- D. guideline

__________ **3.** A(n) ___ is a simple overview of the energy-consuming equipment in a building.
- A. equipment energy review
- B. building survey
- C. energy consumption report
- D. preliminary energy audit

__________ **4.** An energy audit is used to identify ___.
- A. the total energy consumption of a building
- B. any electrical installations that are not up to code
- C. building elements that are not producing proper returns on investments
- D. ways to improve energy efficiency

__________ **5.** Payback periods of ___ are considered acceptable.
- A. three years or less
- B. three to five years
- C. five to ten years
- D. up to ten years

__________ **6.** Complex energy audits may require a(n) ___.
- A. special certification
- B. building permit
- C. team of specialized engineers
- D. additional energy auditor

_______________ **7.** ASHRAE Standard ___, establishes minimum energy efficiency requirements for every type of HVAC equipment and lighting device.

A. 62.1, *Ventilation for Acceptable Indoor Air Quality*
B. 90.1, *Energy Standards for Buildings Except for Low-Rise Residential Buildings*
C. 52, *Method of Testing General Ventilation Air Cleaning Devices for Removal Efficiency by Particle Size*
D. 55, *Thermal Environmental Conditions for Human Occupancy*

_______________ **8.** A(n) ___ is the complete documentation of the energy audit and the results of the data analysis.

A. return on investment summary
B. energy summary
C. energy bill
D. energy audit report

_______________ **9.** Standards may be adopted by federal, state, or local officials as ___ for buildings used by the public.

A. suggestions
B. guidelines
C. maximum requirements
D. minimum requirements

Completion

_______________ **1.** ___ is the ratio of savings versus cost over a specific time period.

_______________ **2.** ___ is the comparison of facilities that have similar characteristics.

_______________ **3.** A(n) ___ is a mandatory rule issued by the local, state, or federal government.

_______________ **4.** A desirable goal for many buildings is to achieve recognition as a green building from the ___.

_______________ **5.** Two methods of measuring effectiveness that are commonly used are called ___ and payback period.

_______________ **6.** A(n) ___ energy audit is the most complex and difficult energy audit to perform.

Section 27.2 Energy-Related Programs and Rating Systems

REVIEW QUESTIONS

Name ______________________________ **Date** ____________________

Multiple Choice

______________ **1.** Energy Star is a program designed to promote energy-efficient products implemented by the U.S. ___ to save energy and reduce power plant greenhouse gas emissions.

A. Environmental Protection Agency (EPA)
B. Green Building Council (USGBC)
C. Department of Energy (DOE)
D. Energy Information Administration (EIA)

______________ **2.** The LEED rating system is broken into four levels, ___, based on LEED points awarded for multiple categories.

A. Certified, Green, Sustainable, and Gold
B. Preliminary, Investment, Residential, and Industrial
C. Certified, Silver, Gold, and Platinum
D. Bronze, Silver, Gold, and Platinum

______________ **3.** LEED certification is granted by the ___.

A. IAMPO
B. USGBC
C. EPA
D. ICC

______________ **4.** An advantage of LEED-certified buildings is ___.

A. poor lighting quality
B. increased total cost of ownership over all building aspects
C. increased solid waste
D. significant savings due to energy efficiency

______________ **5.** Energy Star products normally save between ___ on energy compared to non-Energy Star products.

A. 20% and 30%
B. 35% and 45%
C. 50% and 60%
D. 70% and 80%

_______________ **6.** The LEED Green Building Rating System was developed and is maintained by the ___.
- A. U.S. Green Building Council
- B. ISO Green Building Program
- C. US Environmental Protection Agency
- D. Canadian Green Buiding Council

Completion

_______________ **1.** ___ is a green building rating system that awards points based on specific energy and environmental construction types and activities.

_______________ **2.** ___ is a voluntary labeling program designed to promote energy-efficient products.

_______________ **3.** The Energy Star program includes a component that encourages the use of ___ power sources.

_______________ **4.** The LEED system for evaluating existing buildings is known as ___.

Section 27.3 Utility Bill Analysis Phase

REVIEW QUESTIONS

Name ______________________ **Date** ______________________

Multiple Choice

______________ **1.** The parts of a commercial ___ are electrical consumption, electrical demand, and power factor/fuel recovery.
- A. data analysis
- B. energy retrofit proposal
- C. energy audit
- D. electric utility bill

______________ **2.** Electrical consumption is measured in ___.
- A. kilovolts (kV)
- B. amps (A)
- C. kilowatt-hours (kWh)
- D. gigavolt hours (gVh)

______________ **3.** A(n) ___ is the classification of a customer depending on the type of service and the amount of electricity used.
- A. tier
- B. ratchet clause
- C. energy cost index
- D. exclusion

______________ **4.** A therm is the quantity of gas required to produce ___ Btu.
- A. 100
- B. 1000
- C. 10,000
- D. 100,000

______________ **5.** ___ is an energy source commonly used in commercial buildings.
- A. Gasoline
- B. Fuel oil
- C. Natural gas
- D. Coal

______________ **6.** The ___ method computes electrical demand on a minute-by-minute basis.
- A. fixed interval
- B. sliding window
- C. instantaneous demand
- D. tiered

_______________ **7.** ___ are used to estimate the potential cost savings of a retrofit.
A. Electrical consumption charts
B. Utility rate structures
C. Fuel receipts
D. Utility recovery rates

_______________ **8.** ___ measures the total amount of electricity used during the usual monthly billing period.
A. Electrical demand
B. Energy utilization
C. Energy cost
D. Electrical consumption

_______________ **9.** The time interval used to determine electrical demand is commonly ___.
A. 15 min or 30 min
B. 5 hrs to 6 hrs
C. 12 hrs to 24 hrs
D. 15 days or 30 days

_______________ **10.** True power equals apparent power only when the power factor is ___%.
A. 0
B. 25
C. 50
D. 100

_______________ **11.** Some utilities use a ___ to set the demand charge to the highest monthly interval for up to a year.
A. summer/winter rate
B. tiered rate
C. ratchet clause
D. time-of-day rate

_______________ **12.** With time-of-day rates, the cost of power per unit is most expensive from ___.
A. 7 AM to 5 PM
B. 5 PM to 11 PM
C. 11 PM to 7 AM
D. 12 AM to 12 PM

_______________ **13.** The ___ charge is a fixed monthly fee charged by the gas company for establishing and maintaining service.
A. recovery
B. natural gas
C. customer
D. distribution

______________ **14.** The ___ is the amount of heat energy (in Btu) used in a commercial building divided by the number of square feet in the building.
A. energy cost index (ECI)
B. energy utilization index (EUI)
C. energy demand
D. energy consumption

______________ **15.** Total ___ is calculated by adding the total usage charge and total demand.
A. energy cost
B. electrical power cost
C. monthly electrical bill
D. recovery charge

______________ **16.** The total natural gas bill is found by ___.
A. multiplying the number of therms used by the natural gas cost per therm
B. adding the total usage charge and total demand
C. multiplying the cubic foot quantity by a Btu factor
D. adding the customer charge, distribution charge, environmental recovery cost, and natural gas cost

Completion

______________ **1.** ___ is the total amount of electricity used during a billing period.

______________ **2.** ___ is the highest amount of electricity used during a specific period of time.

______________ **3.** ___ is a measure of electrical efficiency and is commonly expressed as a percentage.

______________ **4.** ___ is the amount of money a utility is permitted to charge to reflect the constantly changing cost of energy.

______________ **5.** A(n) ___ is documentation that permits an electric utility to charge for demand based on the highest amount of electricity used in a 12-month period, not the amount actually measured.

______________ **6.** A(n) ___ rate structure may be used to increase or decrease the cost of power depending on the time of day in which the electricity is used.

______________ **7.** A(n) ___ is the amount of heat energy required to raise the temperature of 1 lb of water 1°F.

______________ **8.** ___ is electrical demand calculated as an average over the time interval.

Natural Gas Bill

_______________ **1.** natural gas cost

_______________ **2.** environmental recovery cost

_______________ **3.** customer charge

_______________ **4.** distribution (delivery) charge

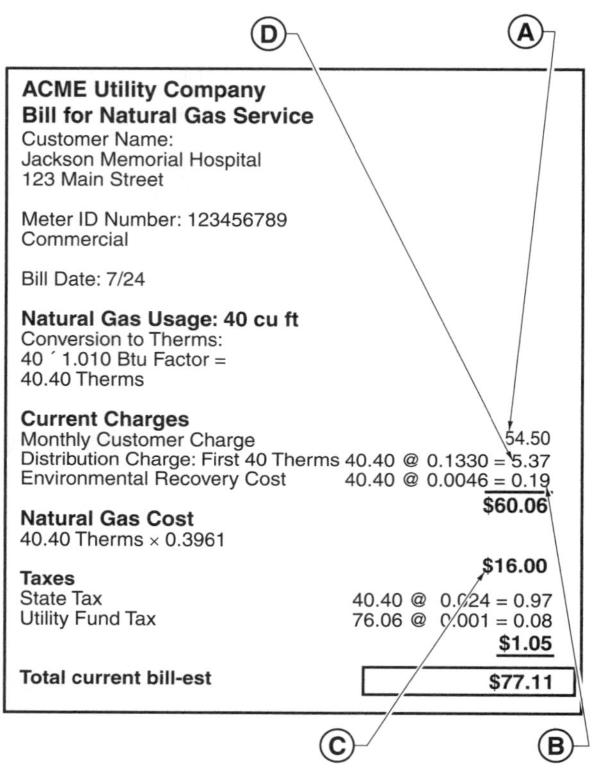

ACME Utility Company
Bill for Natural Gas Service
Customer Name:
Jackson Memorial Hospital
123 Main Street

Meter ID Number: 123456789
Commercial

Bill Date: 7/24

Natural Gas Usage: 40 cu ft
Conversion to Therms:
40 ´ 1.010 Btu Factor =
40.40 Therms

Current Charges
Monthly Customer Charge 54.50
Distribution Charge: First 40 Therms 40.40 @ 0.1330 = 5.37
Environmental Recovery Cost 40.40 @ 0.0046 = 0.19
$60.06

Natural Gas Cost
40.40 Therms × 0.3961
$16.00

Taxes
State Tax 40.40 @ 0.024 = 0.97
Utility Fund Tax 76.06 @ 0.001 = 0.08
$1.05

Total current bill-est $77.11

Section 27.4 Building Survey Phase

REVIEW QUESTIONS

Name ______________________________ Date ________________

Multiple Choice

_______________ **1.** Electric heat values are listed in ___.
A. kilowatts
B. British thermal units
C. therms
D. kilowatt-hours

_______________ **2.** A(n) ___ is used to provide information to the sales engineer and involves reviewing blueprints and interviewing employees.
A. data analysis
B. building survey
C. energy audit report
D. retrofit proposal

_______________ **3.** Observing the dress of building occupants can indicate ___.
A. equipment wear
B. empty refrigerant cylinders
C. energy waste
D. building condition comfort levels

_______________ **4.** Verifying the results of an energy audit includes periodic inspections to ___.
A. ensure maintained efficiency of the new equipment
B. synchronize time schedules
C. identify equipment incompatibilities
D. compare the efficiencies of the old and new equipment

_______________ **5.** During a building survey, the ___ for each load should be included.
A. type of electricity
B. available power sources
C. typical efficiency
D. type of control

_______________ **6.** Large direct expansion refrigeration systems are normally audited by testing for ___ at different load levels.
A. kW/hr
B. kW/ton
C. Btu/hr
D. Btu/ton

Completion

______________ **1.** A(n) ___ is an inventory of the energy-consuming equipment in a commercial building.

______________ **2.** ___ are tested by finding the airflow across the evaporator, wet bulb temperatures entering and leaving the evaporator, and the electrical power demand.

______________ **3.** A reduction in the condensing water supply temperature can lead to large ___.

REVIEW QUESTIONS

Name ____________________ **Date** ____________________

Multiple Choice

__________ **1.** A(n) ___ is a one-page summary of the significant parts of a proposal.
- A. exclusion
- B. executive summary
- C introduction
- D. signature sheet

__________ **2.** The purpose of a written proposal is to ___.
- A. sell the customer an energy retrofit
- B. convince the customer to replace current energy equipment
- C. determine the total cost of utilities for a building
- D. explain the results of an energy audit in simple terms

__________ **3.** The proposal body includes ___.
- A. the drawings of possible retrofit options
- B. a list of equipment to be retrofitted
- C. a clause that guarantees the price given to the customer for a specific period of time
- D. the estimated payback time (in months)

__________ **4.** The ___ protects the contractor from a customer that waits for an extended period of time and expects to get the same price previously quoted.
- A. appendix
- B. executive summary
- C. guarantee clause
- D. exclusion

Completion

__________ **1.** Building automation system ___ are written to sell HVAC equipment and services such as furnaces, water treatment equipment, and preventive maintenance.

__________ **2.** A(n) ___ is an item in a proposal that is the responsibility of the customer and not the contractor.

__________ **3.** A critical part of a(n) ___ is the estimated payback time (in months).

Energy Audits and Utility Structures

ACTIVITIES

Name ______________________ Date ______________

Activity 27-1. Electrical Cost Determination

It is summer and a customer has called indicating that their utility bill is high. The customer wants a breakdown of their utility bill. Use the meter readings and the monthly utility rates to calculate the electrical cost.

__________ **1.** The total usage cost for the month is $___.

__________ **2.** The total demand cost for the month is $___.

__________ **3.** If the fuel recovery charge is $0.02 per kWh, the fuel recovery charge for the month is ___.

__________ **4.** The total electric bill for the month, excluding other taxes and power factor charges, is ___.

__________ **5.** If the building is 10,000 sq ft, the ECI for the month is $___ per sq ft.

MONTHLY UTILITY RATES

KILOWATT DEMAND CHARGE (IN DOLLARS PER kW)		
	SUMMER	WINTER
FOR FIRST 50 kW	$9.62	$8.62
FOR ALL EXCESS OVER 50 kW	$8.34	$7.43

KILOWATT HOUR CHARGE (IN CENTS PER kWh)		
	SUMMER	WINTER
FOR FIRST 40,000 kWh	4.58	4.15
FOR NEXT 60,000 kWh	3.27	2.95
FOR NEXT 200 kWh PER kWd BUT NOT LESS THAN 400,000 kWh	2.86	2.66
FOR NEXT 200 kWh PER kWd	2.182	2.043
FOR ALL EXCESS	0.91	0.82

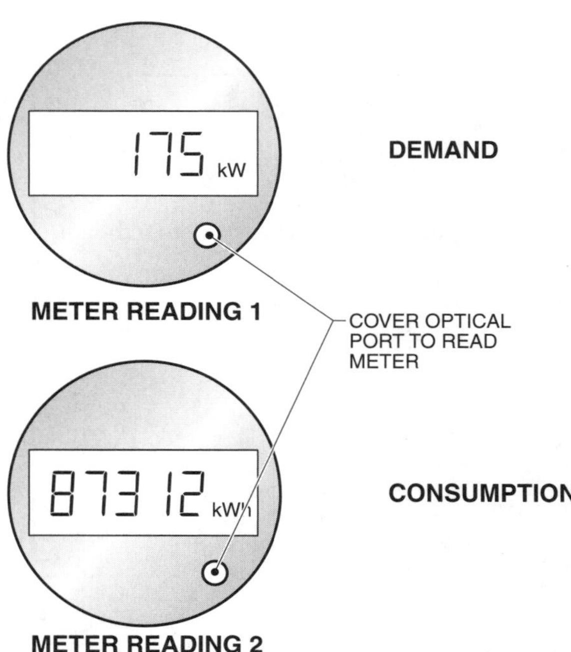

Activity 27-2. Energy Use Analysis

A company has provided last year's utility bills. Plot the company energy use data to help justify expenditures on new mechanical equipment.

MONTHLY ELECTRICAL USAGE/DEMAND		
MONTH	USAGE*	DEMAND†
January	65,500	140
February	67,000	145
March	72,000	155
April	75,000	165
May	80,000	171
June	83,000	177
July	87,000	175
August	89,000	180
September	84,000	165
October	84,000	163
November	72,000	154
December	68,000	152

* in kWh

† in kW

1. Graph the company energy usage data.

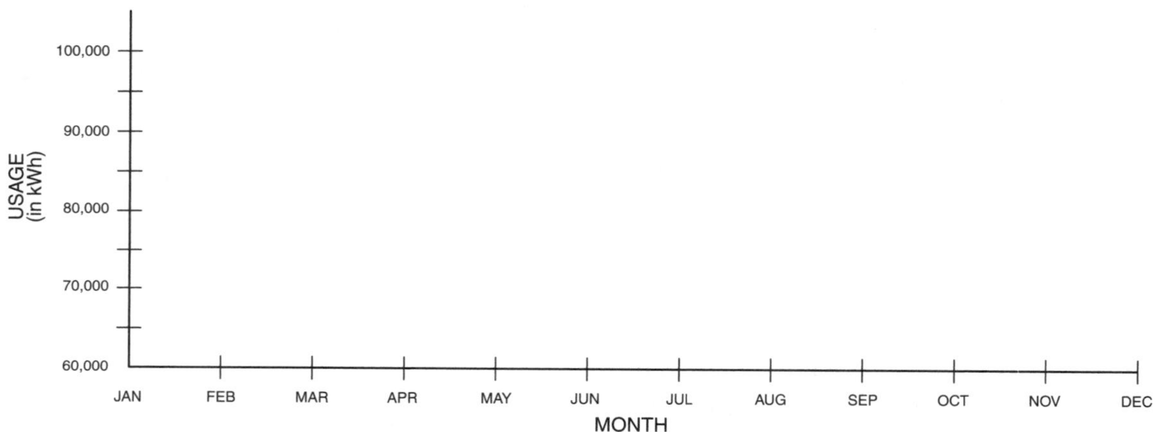

2. Graph the company energy demand data.

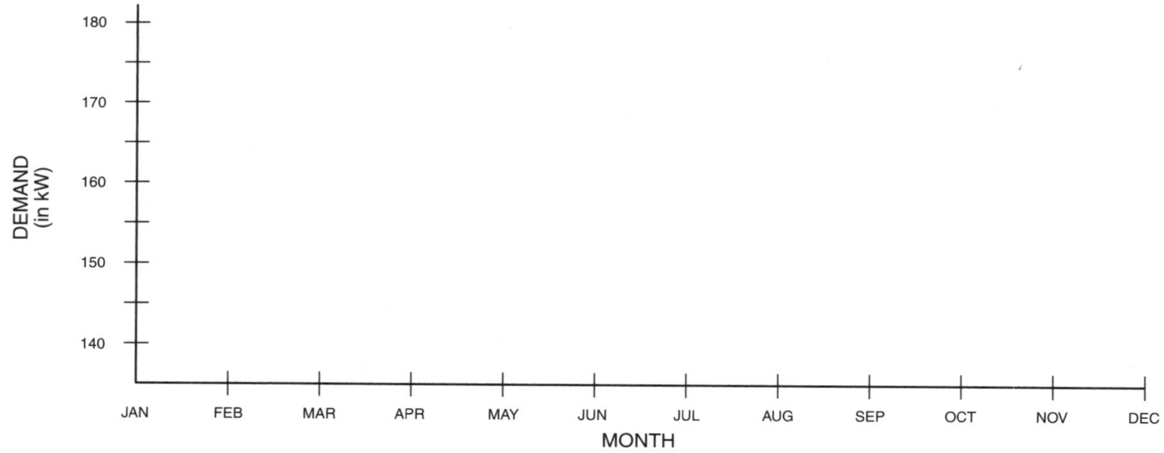

Section 28.1 Human-Computer Interface Troubleshooting

REVIEW QUESTIONS

Name ______________________ Date ______________

Multiple Choice

_______________ 1. Manufacturers provide an override indicator such as a(n) ___ in the building automation system control software that can be accessed from the workstation computer that indicates an override is in effect.
- A. XX
- B. II
- C. O
- D. 1

_______________ 2. If prints or sequences of operation are not available, the service technician may need to rely on ___.
- A. technical support websites or telephone hotlines
- B. the retrofit proposal and inspection records
- C. information from the HCI
- D. judgment and field experience with similar systems

_______________ 3. Few or excessive alarms may be caused by ___.
- A. a lack of trending
- B. a control strategy keeping the system off
- C. the human-computer interface not working
- D. a failure to prioritize alarms properly when setting up the system

Completion

_______________ 1. The ___ of building automation system problems are categorized into human computer interface (HCI), field controller, field input and output, and mechanical equipment.

_______________ 2. A building automation system ___ should be consulted regarding specific procedures for troubleshooting equipment.

_______________ 3. Building automation systems include a feature that allows ___ to review which individuals perform specific tasks such as setting overrides.

Building Automation System Software Indicators

__________ 1. operating indicator

__________ 2. override indicator

__________ 3. shutdown indicator

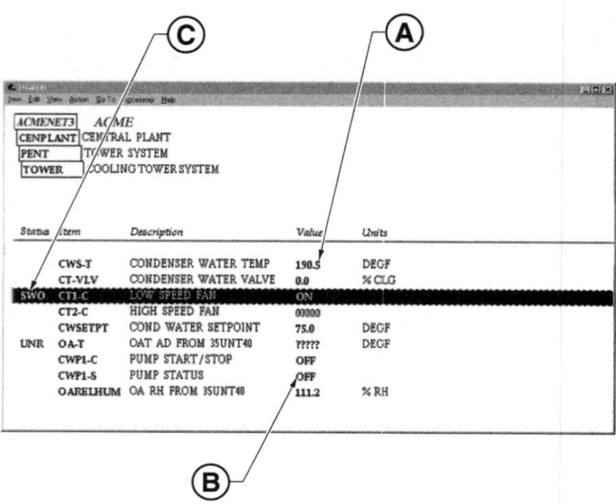

Alarm Parameter Folders

__________ 1. parameter

__________ 2. setpoint

__________ 3. name code

__________ 4. alarmed sensor

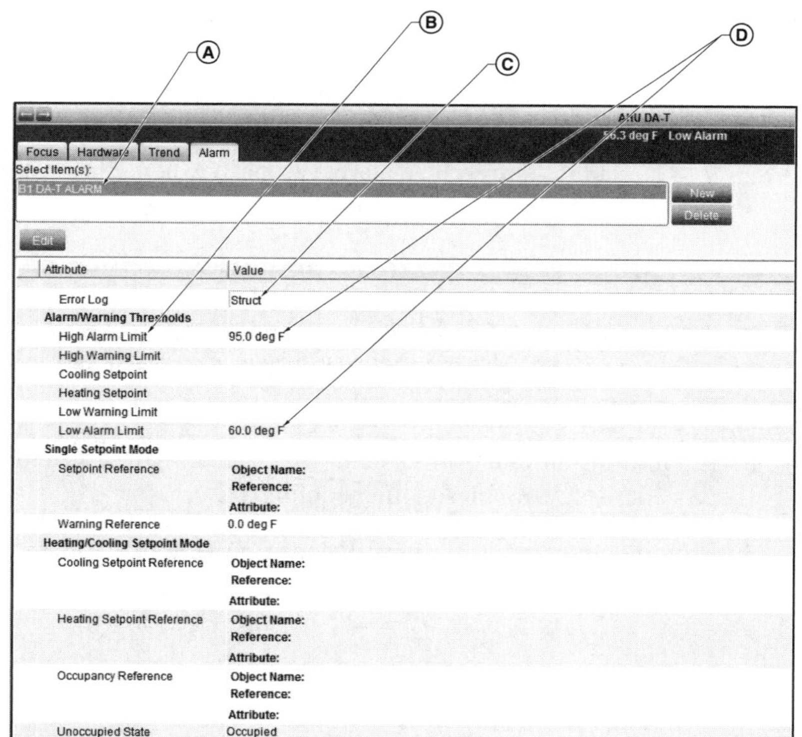

Section 28.2 Field Controller Troubleshooting

REVIEW QUESTIONS

Name ______________________________ **Date** ____________________

Multiple Choice

_______________ **1.** A(n) ___ controller is indicated by an individual controller being off-line.
- A. overloaded
- B. underloaded
- C. constantly cycling
- D. damaged

_______________ **2.** A building space that is too hot may be the result of not enough air volume in the duct supply to the VAV terminal boxes due to ___.
- A. power loss in the controller
- B. the constant cycling of a controlled device
- C. an off-line controller
- D. improper static pressure setpoints

_______________ **3.** A destroyed power supply should be checked using a(n) ___.
- A. digital multimeter
- B. human-computer interface
- C. overload tester
- D. communication bus

Completion

_______________ **1.** Constant ___ of controlled devices is commonly caused by building automation system software that has incorrectly set control loop parameters.

_______________ **2.** The standard static pressure setpoint for a variable air volume system is ___″ wc.

_______________ **3.** The primary cause of excessively hot or cold temperature complaints is having the ___ in the building automation system controller software.

Communication Bus Troubleshooting

______________________ 1. network connection

______________________ 2. network communication module

______________________ 3. controllers

______________________ 4. local communication bus

______________________ 5. HCI

______________________ 6. DMM test 3 problem found at controller 8

______________________ 7. DMM test 1 halfway

______________________ 8. DMM test 2 halfway again

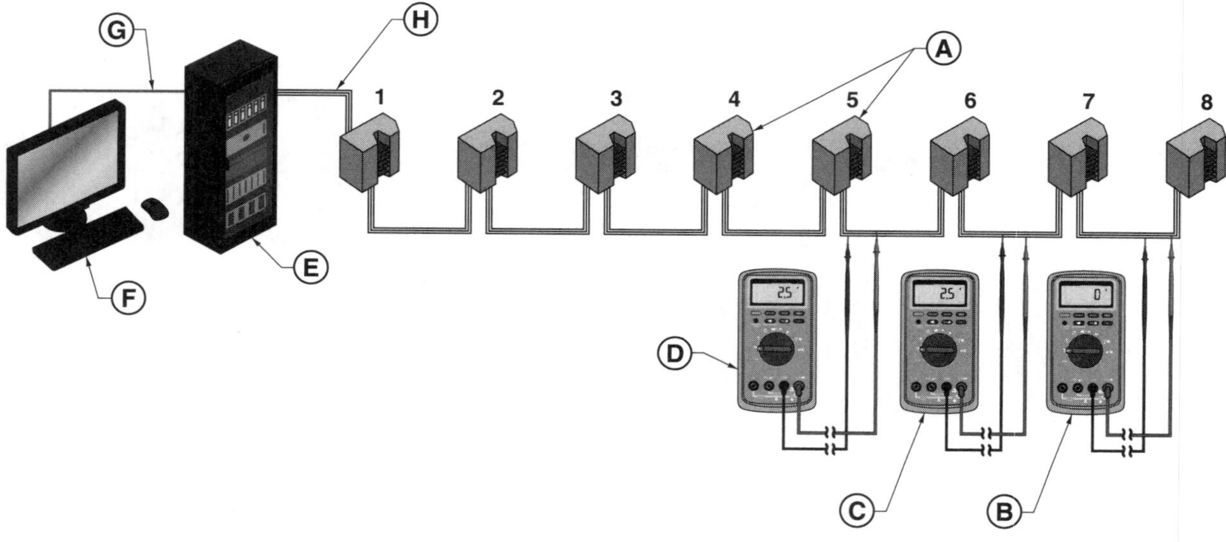

Section 28.3 Field Input and Output Troubleshooting

REVIEW QUESTIONS

Name ______________________________ **Date** ________________

Multiple Choice

______________ **1.** Common ___ input problems include incorrect input setpoint, input failure, and pump status failure.

A. digital
B. mechanical
C. analog
D. automation

______________ **2.** A room temperature sensor that reads properly, then improperly shortly after, may have ___.

A. physical damage to the sensor
B. loose wiring
C. incorrect sensor setpoints
D. multiple or binary inputs

______________ **3.** ___ are commonly used as status switches and thermostats, such as low temperature limit switches, high static pressure limit switches, and fan status switches.

A. Analog inputs
B. Digital inputs
C. Analog outputs
D. Digital outputs

______________ **4.** Analog outputs are almost always ___.

A. 35 mA to 50 mA
B. 40 mA to 60 mA
C. 0 VDC to 10 VDC
D. 10 VDC to 20 VDC

______________ **5.** If a building automation system shows an ON command for a boiler but both the boiler and the indicator lamp are off, the problem is fixed by ___.

A. checking for improper wiring and replacing damaged or burnt wires
B. resetting the BAS output setpoints
C. updating the digital output software
D. replacing the output interface relay

______ **6.** Improper or no sensor readings may be caused by ___.
A. electronic signals that are too high to be read by the DMM
B. physical damage to the sensor or incorrect sensor information
C. improper wiring between the communication bus and network controller
D. a defective DMM

______ **7.** A sensor that reads properly, but in a short period of time reads improperly, may have experienced ___.
A. improper calibration
B. an electrical surge
C. physical damage
D. a short circuit

______ **8.** ___ are used by building automation systems for devices that are ON or OFF and do not require modulation.
A. Digital outputs
B. Analog outputs
C. Digital inputs
D. Analog inputs

Completion

______ **1.** ___ inputs are most commonly used to measure temperature, pressure, or humidity.

______ **2.** To protect sensors, manufacturers sell ___ enclosures that protect a sensor from damage.

______ **3.** Common ___ problems include actuator failure, incorrect wiring, and incorrect software setup.

Section 28.4 Mechanical Equipment Troubleshooting

REVIEW QUESTIONS

Name ______________________________________ Date ____________________

Multiple Choice

______________ 1. When using a variable air volume system, the discharge air temperature should be checked at the air handling unit to ensure it is at ___°F.
- A. 55
- B. 65
- C. 74
- D. 76

______________ 2. After troubleshooting has taken place at the HCI level, field controller level, and input and output level, the fault may be in the ___.
- A. operator
- B. power source
- C. wiring
- D. mechanical equipment

______________ 3. If a VAV terminal box is calling for 500 cfm of air but only receiving 350 cfm, the VAV terminal box is ___.
- A. oversized
- B. undersized
- C. starving for air
- D. underpowered

______________ 4. If the chilled water supply temperature is fluctuating from 42°F to 48°F at the chiller, the valves and chiller are ___.
- A. hunting
- B. starving
- C. setup properly
- D. in a lead/lag system

______________ 5. A ___ can indicate that the HVAC equipment is not capable of satisfying demand.
- A. modulating damper
- B. half-open valve
- C. modulating valve
- D. wide-open valve

Completion

______________________ 1. If a mechanical equipment problem is suspected, the mechanical equipment should be checked using the ___ or portable operator terminal to determine any problems with the mechanical equipment.

______________________ 2. When a VAV terminal box is starved for air despite having a 100% open damper, ___ the static setpoint to allow a greater volume of air into the duct.

______________________ 3. When two chillers are operating in a lead/lag system and the leading chiller has a fault that requires the second chiller to run, the cause is likely to be ___.

Building Automation System Troubleshooting

ACTIVITIES

Name ____ Date ____________________________

Activity 28-1. Sensor Troubleshooting

While checking a building automation system operator workstation screen, a system sensor appears to have a problem. Use the operator workstation display to answer the questions.

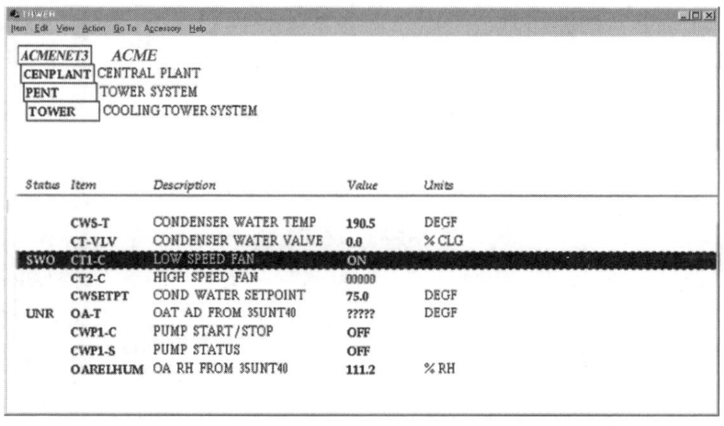

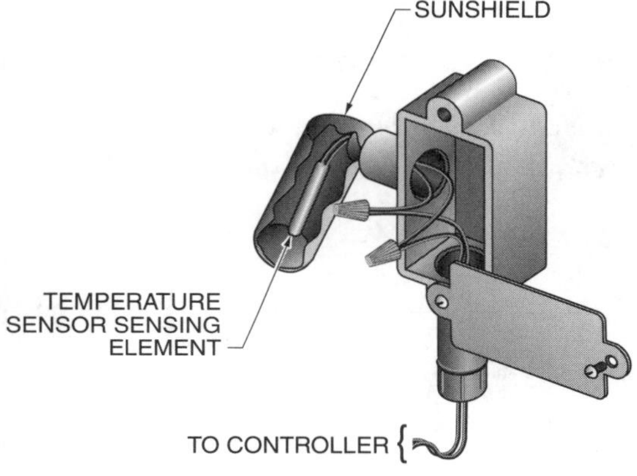

OPERATOR WORKSTATION DISPLAY

____________________ **1.** The ___ sensor (item) has question marks in the value column.

____________________ **2.** The three letters that are listed in the status column of the sensor are ___.

____________________ **3.** What might this mean?

4. What is the next logical step since it is suspected that this sensor has a problem?

____

____

A DMM is used to test the sensor after determining its location in the system. The sensor is listed as a PTC sensor that has a nominal value of 1000 Ω at 70°F and a coefficient of 2.2 Ω per °F. The outside air temperature is 85°F.

____________________ **5.** The sensor resistance should be ___ Ω.

The sensor is disconnected from the building automation system controller, and a DMM set to measure resistance is used to check the sensor. The sensor resistance is 3 Ω.

6. What is the problem?

____

____

7. What should be done?

____

____

Activity 28-2. Building Automation System Software Troubleshooting

While checking a building automation system operator workstation screen, it is determined that a tower low-speed fan is ON when it should be OFF. Use the operator workstation display to answer the questions.

________________ **1.** The tower current value listed on the display is ___.

________________ **2.** The tower listed command priority is ___.

________________ **3.** The tower commanded feature is ___.

________________ **4.** The tower communication status is ___.

________________ **5.** The listed S/W override (in the software) is ___.

________________ **6.** Does this indicate that the fan has been overridden by an operator?

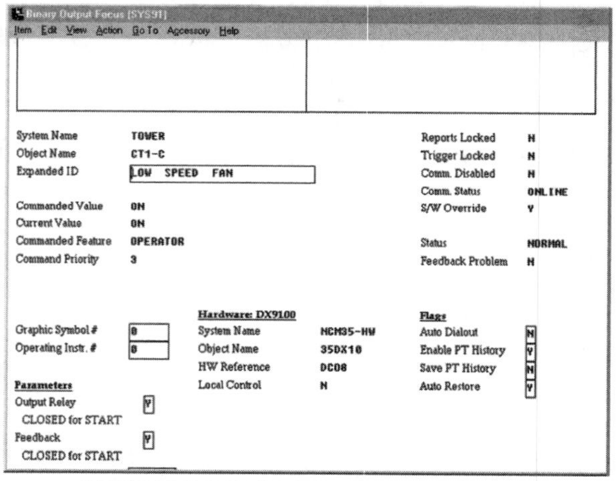

Binary Output Focus [SYS91]
Item Edit View Action Go To Accessory Help

System Name TOWER
Object Name CT1-C
Expanded ID LOW SPEED FAN
Commanded Value ON
Current Value ON
Commanded Feature OPERATOR
Command Priority 3
Reports Locked N
Trigger Locked N
Comm. Disabled N
Comm. Status ONLINE
S/W Override Y
Status NORMAL
Feedback Problem N
Graphic Symbol # 0
Operating Instr. # 0
Hardware: DX9100
System Name NCM35-HW
Object Name 35DX10
HW Reference DC08
Local Control N
Flags
Auto Dialout N
Enable PT History Y
Save PT History N
Auto Restore Y
Parameters
Output Relay Y
CLOSED for START
Feedback Y
CLOSED for START

OPERATOR WORKSTATION DISPLAY

7. What should be done to correct the problem?

____

____

____

Activity 28-3. Troubleshooting a Constantly Cycling Control Device

A maintenance technician has noticed that a heating valve is constantly cycling open and closed. A laptop computer is used to open the software folder for the controller. Use the controller software folder information to answer the questions.

____________________ **1.** The controller Occ Htg Setpt is ___.

____________________ **2.** The controller Htg Prop Band is ___.

____________________ **3.** The controller Htg Integ Time is ___.

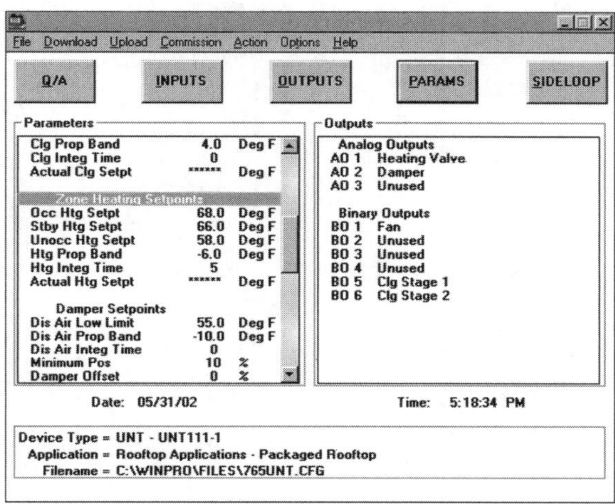

CONTROLLER SOFTWARE FOLDER

The original controller software that was used when the heating loop worked properly was found. The original values were 68°F for Occ Htg Setpt, –10°F for Htg Prop Band, and 20 for Htg Integ Time.

4. What should be done?

____

____

5. What should be done to ensure proper operation?

____

____

Section 29.1 History of Interoperability

REVIEW QUESTIONS

Name ______________________________ Date ____________________

Multiple Choice

_______________ **1.** A(n) ___ standard is published by an established governing body and adopted by a national or international standards group.

A. controller
B. closed system
C. open protocol
D. intranet

_______________ **2.** ___ is the ability of building automation systems from different vendors to share data, control, and information functions across a common network.

A. Interchangeability
B. Integration
C. Interoperability
D. Routing

_______________ **3.** Public facilities are legally required to participate in a(n) ___ process for building automation systems.

A. vendor-driven bidding
B. custom-system presentation
C. open-bid
D. open forum

_______________ **4.** A(n) ___ is a single computer loaded with special software that permits the viewing of different manufacturers' software.

A. computer gateway
B. open protocol
C. human-computer interface
D. physical layer device

_______________ **5.** The ___ layer is used by programs to access network services.

A. physical
B. application
C. BACnet
D. OSI

______ **6.** The ___ model describes network protocols and communication.
- A. ISO
- B. OSI
- C. open protocol
- D. LonWorks

______ **7.** In the OSI model, the ___ describe how the applications relate to the network.
- A. upper layers
- B. lower layers
- C. application specifications
- D. data link layers

Completion

______ **1.** The ___ layer of the open systems integration (OSI) seven-layer model defines the layout of devices such as hubs, repeaters, and wiring.

______ **2.** ___ is the ability of one device to be substituted for another device and for the system to operate correctly.

______ **3.** In the OSI model the ___ layers describe the aspects of how communications are sent from one computer to another.

______ **4.** A(n) ___ is any publicly published protocol.

______ **5.** The ___ layer of the OSI handles data routing across the different segments of the network.

REVIEW QUESTIONS

Name ______________________________ Date ____________________

Multiple Choice

_______________ **1.** A ___ is an analog input, analog output, digital input, or digital output, each of which has different characteristics.
- A. map
- B. point
- C. protocol
- D. network

_______________ **2.** The ___ object represents sensor input.
- A. trend log
- B. notification class
- C. analog output
- D. analog input

_______________ **3.** The five categories of ___ are data sharing, alarm and event management, scheduling, trending, and device and network management.
- A. BACnet local area networks (LANs)
- B. BACnet broadcast management devices (BBMDs)
- C. protocol implementation conformance statements (PICS)
- D. BACnet interoperability building blocks (BIBBs)

_______________ **4.** The types of local area network (LAN) technology are master-slave/token passing (MS/TP) and ___.
- A. ARCNET
- B. Point-to-Point (PTP)
- C. LonTalk
- D. BACnet/IP

_______________ **5.** The ___ feature of a network tool is used to automatically locate and identify BACnet nodes on a network.
- A. automated discovery
- B. exposing
- C. search and identify
- D. intrinsic location

_______________ **6.** The ___ are used to give each device object a unique identifier so that it may be used on the network.
A. device sequences
B. identification codes
C. object instance numbers
D. device instance numbers

_______________ **7.** A ___ node is a node that does not participate in token passing but only responds when a specific request is sent.
A. slave
B. backup
C. master
D. fail-safe

_______________ **8.** ___ tools include a network analyzer and network diagnostic tools at the device.
A. Hardware
B. Software
C. Commissioning
D. Retrofit

_______________ **9.** The ___ was created to certify products as meeting the BACnet standard.
A. Occupational Standards and Health Administration (OSHA)
B. American Society of Heating, Refrigerating and Air-Conditioning Engineers (ASHRAE)
C. BACnet Testing Laboratory (BTL)
D. International Standards Organization (ISO)

Completion

_______________ **1.** ___ developed and maintains the Building Automation and Control Networks (BACnet).

_______________ **2.** The BACnet standard has approximately ___ standard object types.

_______________ **3.** A(n) ___ is a detailed description for a specific manufacturer's BACnet device that states its BACnet capabilities.

_______________ **4.** ___ types of local area network (LAN) technology can be used to transport BACnet messages from node to node.

_______________ **5.** A(n) ___ is a device used to forward broadcast messages between IP subnets that contain BACnet/IP nodes.

_______________ **6.** ___ is a feature of certain BACnet objects in which the alarm or event originates from the object.

_______________ **7.** ___ nodes can be broken down into master nodes and slave nodes.

_______________ **8.** ___ is one of the most popular protocol standards for building automation systems.

Section 29.3 LonWorks Systems

REVIEW QUESTIONS

Name ______________________________ Date ____________________

Multiple Choice

______________ **1.** The ___ contains a whole system-on-a-chip with multiple processors and read-write and read-only memory (RAM and ROM), communications, and input/output (I/O) subsystems.
A. peer-to-peer network
B. router
C. repeater
D. Neuron chip

______________ **2.** In the LonWorks infrastructure, ___ provide error checking of each message.
A. routers
B. repeaters
C. Neuron chips
D. function profiles

______________ **3.** LonWorks systems use ___ nodes in customized sequences such as large complex air handlers, boilers, and chiller plant control.
A. application-specific
B. building block
C. free-programmable
D. interoperable

______________ **4.** Two major node types are application-specific nodes and ___ nodes.
A. change-of-valve
B. interoperability
C. free-programmable
D. performance binding

______________ **5.** ___ are used to define network variable behavior.
A. Winks
B. Heartbeats
C. Bindings
D. Configuration properties

_______________ **6.** A(n) ___ is a data item that a device application program receives from or sends to other devices on a network.
- A. network variable
- B. node
- C. output
- D. point

_______________ **7.** One unique characteristic of LonWorks is the implementation of a(n) ___ on every LonWorks controller.
- A. service pin
- B. Neuron chip
- C. auxiliary device
- D. control sensor

_______________ **8.** One disadvantage of LonWorks is that the use of ___ may limit controller capabilities.
- A. scaled systems
- B. Neuron chips
- C. free-programmable nodes
- D. communication cards

_______________ **9.** Network ___ are used to connect two network variables so they can share data.
- A. links
- B. configurations
- C. winks
- D. bindings

_______________ **10.** The ___ is the top level of a LonWorks network address and is used to identify the installation.
- A. node number
- B. domain
- C. subnet
- D. network variable input (NVI)

Completion

_______________ **1.** Data received from another device is known as ___.

_______________ **2.** A(n) ___ contains a whole system-on-a-chip with multiple processors, read-write and read-only memory, communications, and input/output subsystems.

_______________ **3.** ___ devices are separated into groups according to their functions.

_______________ **4.** A(n) ___ is the length of time that a value is repeated on a network.

_______________ **5.** ___ is the open protocol standard used in LonWorks control networks.

Section 29.4 Point Mapping and LonWorks Network Testing

REVIEW QUESTIONS

Name ______________________________ **Date** ____________________

Multiple Choice

________________ **1.** In the building automation system installation process, the field devices and field points are identified to a ___.
- A. point map
- B. head-end supervisory device
- C. field controller
- D. hub

________________ **2.** Network performance must be ___ to avoid data collisions, corrupted data, and slow network operation.
- A. sequenced
- B. optimized
- C. backed-up
- D. private and secure

________________ **3.** A(n) ___ may be needed in the event of a major network problem.
- A. adaptive tuning
- B. resynchronization
- C. self-tuning controller
- D. control-loop tuning

________________ **4.** The ___ feature consists of a software utility that searches the selected network, interrogates, and automatically locates the field controllers on the network.
- A. adaptive tuning
- B. intrinsic protocol
- C. autodiscover
- D. self-correct

Completion

________________ **1.** In a LonWorks system, the file type may be a(n) ___.

________________ **2.** The control sequences can be verified and tested by overriding variables in the ___ controller.

________________ **3.** A(n) ___ is a management tool that updates all nodes and databases with the most recent changes.

Section 29.5 Advanced HVAC Control Technologies

REVIEW QUESTIONS

Name ______________________________ Date ____________________

Multiple Choice

________________ **1.** ___ is the ability of a controller to self-diagnose and self-correct a system in order to correct various problems.

A. Adaptive tuning
B. Resynchronization
C. Remote calibration
D. Protocol conformance

________________ **2.** ___ communication is the use of a high-frequency electronic signal to communicate between control system components.

A. Infrared
B. Intranet
C. PC-based
D. Radio frequency

________________ **3.** Mechanical equipment wear, lack of maintenance, and incorrect initial controller programming can be the causes of improper ___.

A. system operation
B. wearing
C. synchronization
D. tuning

________________ **4.** ___ result in increased energy efficiency and system accuracy.

A. Radio frequency communication devices
B. Automated discoveries
C. Self-tuning controllers
D. Software plug-ins

________________ **5.** Hard-wiring problems are eliminated and quick reconfiguration of a BAS is made possible by using ___.

A. local operating networks
B. wireless networks
C. local area networks
D. radio frequency communication

______________ **6.** A drawback of radio frequency technology is that ___ occurs between radio frequency devices.
- A. signal dropping
- B. interference
- C. resynchronization
- D. winking

Completion

______________ **1.** Advanced HVAC control technologies help reduce control system ___ and improve control system accuracy.

______________ **2.** ___ controllers release technicians from tuning control loops.

______________ **3.** Poor system control is the result of insufficient time spent ___ a system.

______________ **4.** ___ may be used to correct for simple errors in installation or application.

Building Automation System Interoperability: Advanced Technologies

ACTIVITIES

Name ______________________________ **Date** ______________

PROTOCOL IMPLEMENTATION CONFORMANCE STATEMENT

Table A: BACnet Protocol Implementation Conformance Statement

Vendor Name:	ABC Controls
Product Name:	NC35
Model Numbers:	NC35-0, NC35-1

Table B: BACnet Conformance Class Supported

Class 1	□	Class 4	■
Class 2	□	Class 5	□
Class 3	□	Class 6	□

Table C: BACnet Functional Groups Supported

Clock	■	Files	□
HHWS	□	Reinitialize	■
PCWS	□	Virtual Operator Interface	□
Event Initiation	■	Virtual Terminal	□
Event Response	■	Device Communications	□
COV Event Initiation	□	Time Master	□
COV Event Response	□		

Product Description

The ABC NC35 supervisory controller is designed to manage a small building or campus of buildings. The NC35 efficiently supervises the networking of application specific controllers (ASCs) and provides facility management features including weekly scheduling, alarm management, optimal start, and trending.

Facility personnel can review the system status and modify control parameters for the NC35 supervisory controller and its associated ASCs using a VT100 terminal or a graphical workstation.

With the addition of a network card, multiple NC35's can communicate over an Ethernet network, providing increased functionality for complex systems.

Table D: BACnet Standard Application Services Supported

Application Service	Initiates Requests	Executes Requests
Acknowledge Alarm	■	■
Confirmed COV Notification	□	□
Confirmed Event Notification	■	■
Get Alarm Summary	□	■
Get Enrollment Summary	□	■
Get Event Information	□	■
Subscribe COV	□	□
Subscribe COV Property	□	□
Unconfirmed COV Notification	■	■
Unconfirmed Event Notification	■	■
Atomic Read File	□	□
Atomic Write File	□	□
Add List Element	■	■
Remove List Element	■	■
Create Object	■	■
Delete Object	■	■
Read Property	■	■
Read Property Conditional	□	□
Read Property Multiple	■	■

Application Service	Initiates Requests	Executes Requests
Read Range Service	□	□
Write Property	■	■
Write Property Multiple	■	■
Device Communication Control	□	□
Confirmed Private Transfer	■	■
Unconfirmed Private Transfer	■	■
Reinitialize Device	□	■
Confirmed Text Message	□	□
Unconfirmed Text Message	□	□
UTC Time Synchronization	■	■
Time Synchronization	□	□
Who-Has	■	■
I-Have	■	■
Who-Is	■	■
I-Am	■	■
VT-Open	□	□
VT-Close	□	□
VT-Data	□	□
Authenticate	□	□
Request Key	□	□

Activity 29-1. BACnet Protocol

A hospital is considering the replacement of an old DDC system. An information sheet (PICS document) has been provided describing the BACnet capabilities of a possible new controller. Use the PICS document to answer the questions.

______________ **1.** The controller manufacturer is ___.

______________ **2.** The controller (product) name is ___.

______________ **3.** Can a VT100 terminal be used as an operator interface?

______________ **4.** The BACnet conformance class that is supported is ___.

5. List the BACnet functional groups that are supported.

____

____

______________ **6.** Does the controller support authenticate?

______________ **7.** Does the controller support acknowledge alarm?

______________ **8.** Does the controller support who-has?

______________ **9.** Does the controller support VT-open?

______________ **10.** Does the controller support read property?

______________ **11.** Does the controller support subscribe COV?

______________ **12.** Does the controller support write property?

______________ **13.** Does the controller support read range service?

______________ **14.** Is the PICS document valid for other controllers manufactured by ABC Controls?

______________ **15.** Could future upgrades of the NC35 controller add more features?

______________ **16.** A facility wishes to integrate a chiller control system that is listed as BACnet compatible. Does this automatically mean that all of the objects and features can be shared between the NC35 and the chiller controller?

______________ **17.** What document should be obtained from the chiller controller manufacturer?

______________ **18.** A facility specification includes the phrase, "Controllers must be BACnet compatible." Is this wording adequate?

Section 30.1 Building Commissioning

REVIEW QUESTIONS

Name ______________________ Date ______________

Multiple Choice

______ **1.** ___ brings the performance of an existing building up to its original intended design.
A. Secondary commissioning
B. Recommissioning
C. Ongoing commissioning
D. Retro-commissioning

______ **2.** ___ is the process of balancing and adjusting an HVAC system for maximum efficiency.
A. Retrofitting
B. Commissioning
C. Tuning
D. Calibration

______ **3.** In the commissioning process, buildings may be categorized by ___.
A. function
B. number of floors
C. energy efficiency level
D. size in square feet

______ **4.** The ___ is responsible for implementing and planning the commissioning process.
A. commissioning authority
B. controls technician
C. owner
D. mechanical contractor

______ **5.** ___ can help reverse unintended equipment and sequence problems caused by an operational staff's attempt to fix immediate problems.
A. Existing commissioning
B. Ongoing commissioning
C. Retro-commissioning
D. Recommissioning

______ **6.** Floor plan changes that change the water flow or airflow patterns require ___.
A. recommissioning
B. ongoing commissioning
C. reactive maintenance
D. reliability-centered maintenance

Completion

______________ **1.** Building commissioning is performed by a(n) ___.

______________ **2.** ___ is a regularly documented process that brings the performance of an existing building up to its original intended design.

______________ **3.** ___ commissioning is needed because the operation of the mechanical system changes over time.

______________ **4.** The first step in the commissioning process is to perform a(n) ___.

______________ **5.** A responsibility of the ___ is to write a clear statement of the design intent for each building system.

Section 30.2 Controls Technicians and the Commissioning Process

REVIEW QUESTIONS

Name ______________________________ Date ____________________

Multiple Choice

_______________ **1.** Any HVAC or other systems found to be operating improperly must be recorded in the commissioning report and reported to the ___ for correction.
A. subcontractor
B. controls technician
C. CA
D. mechanical contractor

_______________ **2.** One area addressed in building commissioning is the ___ of a facility.
A. water supply
B. fuel supply
C. operations and maintenance (O&M)
D. indoor lighting

_______________ **3.** A ___ maintenance program is a maintenance program in which worn equipment is repaired or replaced as it fails.
A. reactive
B. preventive
C. predictive
D. reliability-centered

_______________ **4.** A ___ maintenance program is a maintenance program in which the current condition of equipment is tested to determine what, if any, maintenance should be performed.
A. reactive
B. preventive
C. predictive
D. reliability-centered

_______________ **5.** Control cabinets have a pocket for a copy of the ___.
A. maintenance log
B. equipment function instructions
C. control drawings
D. commissioning goals

______________ **6.** As part of the ongoing commissioning process, it is important to check the ___ of each VAV terminal box against a measuring device.
A. airflow volume values
B. resistance
C. valve position
D. offset

______________ **7.** ___ measure the condition of equipment and systems in a building.
A. Equipment metering practices
B. Equipment diagnostics
C. Predictive maintenance programs
D. Reactive maintenance programs

Completion

______________ **1.** A(n) ___ maintenance program is a maintenance program in which routine maintenance is scheduled to reduce failure and prolong equipment life.

______________ **2.** The different types of maintenance programs include ___, preventive, predictive, and reliability-centered maintenance programs.

______________ **3.** A(n) ___ maintenance program recognizes that it is often more cost-effective to repair or replace specific inexpensive or unimportant parts only as needed.

Building Commissioning

ACTIVITIES

Name ______________________________ **Date** ____________________

Activity 30-1. Building System Checks

The building commissioning process includes checks on major systems such as lighting, HVAC, and control systems. Commissioning checks are normally performed by having the HVAC or other contractor operate the equipment while the commissioning authority is present. Today, this is normally done by using the building automation system to override the equipment to the desired state.

Note: The following activities should only be completed with permission and under the supervision of authorized personnel. Proper safety equipment must be used and the proper safety procedures must be followed. Measurements should only be taken by properly trained and authorized individuals.

1. Use the lighting checklist to access the operation of the system.

LIGHTING CHECKLIST	
	Specifications
Location	
Lighting Type	
Fixture Manufacturer & Model Number	
Lamp Types	
Lamp Manufacturer	
Total Wattage	
Total Area Served	
Illumination (W/ft²)	

Control System	

Operating Schedule	

Notes	

2. Use the boiler checklist to access the operation of the system.

BOILER CHECKLIST	
	Specifications
Location	
Manufacturer & Model Number	
Serial Number	
Type	
Fuel	
Number of Passes	
Ignition Type	
Burner Control	
Voltage/Phase/Frequency	

	Design	Actual
Operating Pressure		
Operating Temperature		
Temperature In		
Temperature Out		
Number of Safety Valves		
Safety Valve Size		
Safety Valve Setting		
High Limit Setting		
Operating Control Setting		
High Fire Setpoint		
Low Fire Setpoint		
Voltage (T1-T2; T2-T3; T3-T1)		
Current (T1; T2; T3)		
Power Draw		
Power Factor		
Draft Fan Voltage		
Draft Fan Current		
Manifold Pressure		
Output (MBtu/h, kW)		
Safety Controls Checked		

Notes

3. Use the chiller checklist to access the operation of the system.

CHILLER CHECKLIST	
	Specifications
Location	
Manufacturer & Model Number	
Serial Number	
Type	
Capacity	
Refrigerant	

		Design	**Actual**
EVAPORATOR	Pressure		
	Temperature		
	Water Pressure In		
	Water Pressure Out		
	Water Temperature In		
	Water Temperature Out		
	Flow Rate		
CONDENSER	Pressure		
	Temperature		
	Water Pressure In		
	Water Pressure Out		
	Water Temperature In		
	Water Temperature Out		
	Flow Rate		
COMPRESSOR	Suction Pressure		
	Suction Temperature		
	Discharge Pressure		
	Discharge Temperature		
	Oil Pressure		
	Oil Temperature		
	Voltage (T1-T2; T2-T3; T3-T1)		
	Current (T1; T2; T3)		
	Power Draw		
	Power Factor		

Notes

4. Use the air handling unit checklist to access the operation of the system.

AIR HANDLING UNIT CHECKLIST		
		Specifications
	Location	
	Manufacturer & Model Number	
	Serial Number	
	Type	
	Arrangement & Class	
	Discharge	
	Sheave Make & Sizes	
	Number of Belts, Make, and Size	
	Belt Drive Center Distance	
MOTOR	Make & Frame	
MOTOR	HP/Voltage/Phase/Frequency	
MOTOR	Full Load Current	
MOTOR	Sheave Make & Sizes	

	Design	Actual
Total Airflow		
Total Static Pressure		
Fan Speed		
Motor Brake HP		
Outdoor Airflow		
Return Airflow		
Motor Speed		
Motor Voltage		
Motor Current		
Motor Power Draw		
Motor Power Factor		
Discharge Static Pressure		
Suction Static Pressure		
Reheat Coil Static Pressure Drop		
Cooling Coil Static Pressure Drop		
Preheat Coil Static Pressure Drop		
Filter Static Pressure Drop		
Vortex Damper Position		
Outdoor Air Damper Position		
Return Air Damper Position		

Notes	

5. Use the air-cooled condenser checklist to access the operation of the system.

AIR-COOLED CONDENSER CHECKLIST	
	Specifications
Location	
Manufacturer & Model Number	
Serial Number	
Type	
Capacity	
Refrigerant	
Control Type	

	Design	**Actual**
Condenser Pressure		
Condenser Temperature		
Air Dry Bulb Temperature In		
Air Dry Bulb Temperature Out		
Fan Speed		
Airflow		
Duct Inlet Static Pressure		
Duct Outlet Static Pressure		
Damper Position		
Fan Cycle		
Motor HP		
Voltage (T1-T2; T2-T3; T3-T1)		
Current (T1; T2; T3)		
Power Draw		
Power Factor		

Notes

6. Use the electric duct heater checklist to access the operation of the system.

ELECTRIC DUCT HEATER CHECKLIST	
	Specifications
Location	
Manufacturer & Model Number	
Serial Number	
Power	
Stages	
Voltage	
Face Area	
Minimum Air Velocity	

	Design	**Actual**
Velocity		
Airflow		
Air Temperature In		
Air Temperature Out		
Voltage (T1-T2; T2-T3; T3-T1)		
Current (T1; T2; T3)		

Notes

7. Use the pump checklist to access the operation of the system.

PUMP CHECKLIST	
	Specifications
Service	
Location	
Manufacturer & Model Number	
Serial Number	
Design Flow Rate	
Design Head	
Pump Speed	
Impeller Diameter	
Motor Manufacturer	
Motor Frame	
Motor HP	
Motor Speed	
Voltage/Phase/Frequency	
Motor Full Load Current	
Pump Off Pressure	
Valve Shut Difference	
Actual Impeller Diameter	
Valve Open Differential Pressure	
Valve Open Flow Rate	

	Design	**Actual**
Final Suction Pressure		
Final Discharge Pressure		
Final Differential		
Final Flow Rate		
Voltage (T1-T2; T2-T3; T3-T1)		
Current (T1; T2; T3)		
Power Draw		
Power Factor		

Notes	

APPENDIX

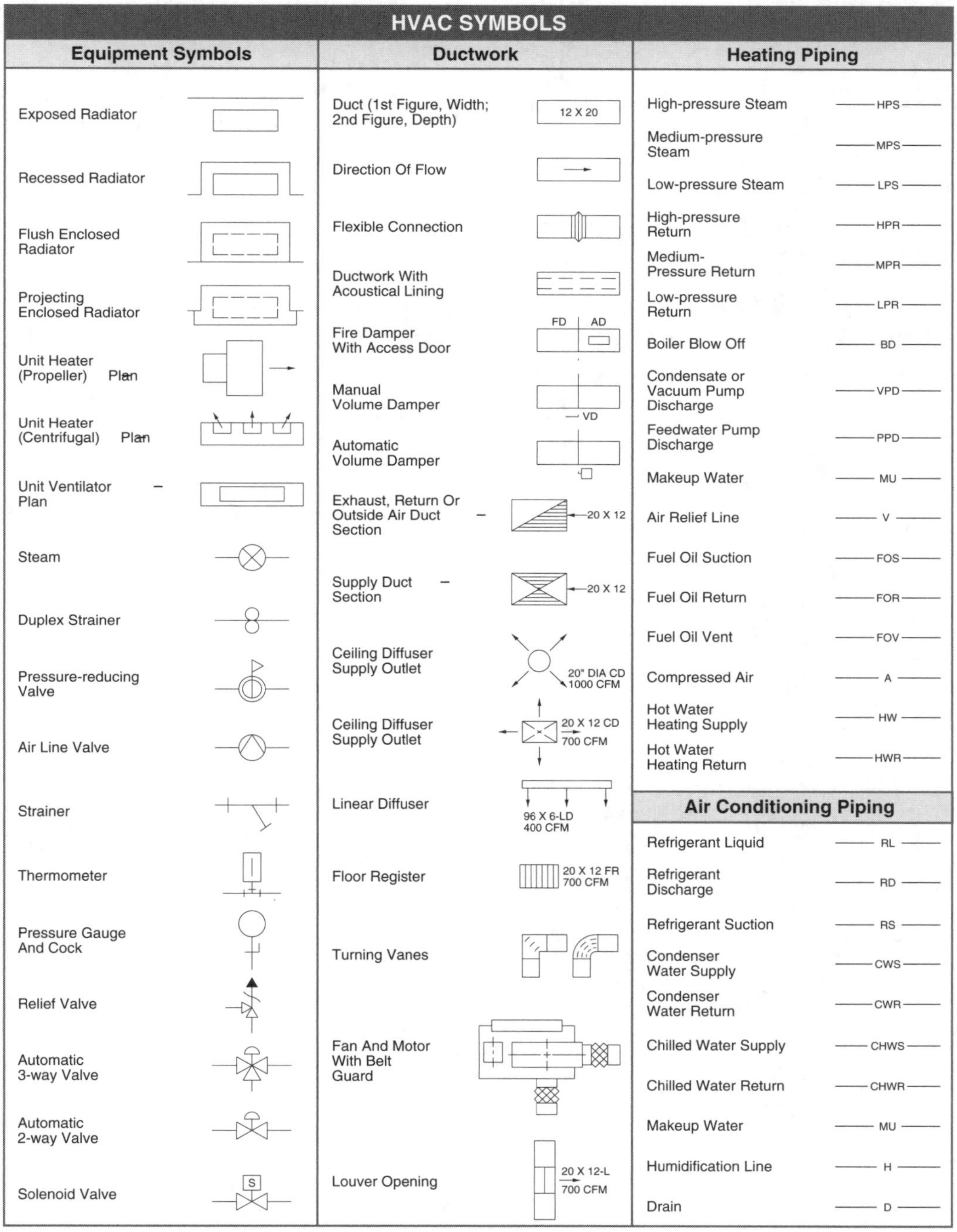

HVAC SYMBOLS

Equipment Symbols

Symbol Name	Symbol
Exposed Radiator	
Recessed Radiator	
Flush Enclosed Radiator	
Projecting Enclosed Radiator	
Unit Heater (Propeller) Plan	
Unit Heater (Centrifugal) Plan	
Unit Ventilator – Plan	
Steam	
Duplex Strainer	
Pressure-reducing Valve	
Air Line Valve	
Strainer	
Thermometer	
Pressure Gauge And Cock	
Relief Valve	
Automatic 3-way Valve	
Automatic 2-way Valve	
Solenoid Valve	S

Ductwork

Symbol Name	Symbol
Duct (1st Figure, Width; 2nd Figure, Depth)	12 X 20
Direction Of Flow	
Flexible Connection	
Ductwork With Acoustical Lining	
Fire Damper With Access Door	FD AD
Manual Volume Damper	VD
Automatic Volume Damper	
Exhaust, Return Or Outside Air Duct – Section	20 X 12
Supply Duct – Section	20 X 12
Ceiling Diffuser Supply Outlet	20" DIA CD 1000 CFM
Ceiling Diffuser Supply Outlet	20 X 12 CD 700 CFM
Linear Diffuser	96 X 6-LD 400 CFM
Floor Register	20 X 12 FR 700 CFM
Turning Vanes	
Fan And Motor With Belt Guard	
Louver Opening	20 X 12-L 700 CFM

Heating Piping

Symbol Name	Symbol
High-pressure Steam	HPS
Medium-pressure Steam	MPS
Low-pressure Steam	LPS
High-pressure Return	HPR
Medium-Pressure Return	MPR
Low-pressure Return	LPR
Boiler Blow Off	BD
Condensate or Vacuum Pump Discharge	VPD
Feedwater Pump Discharge	PPD
Makeup Water	MU
Air Relief Line	V
Fuel Oil Suction	FOS
Fuel Oil Return	FOR
Fuel Oil Vent	FOV
Compressed Air	A
Hot Water Heating Supply	HW
Hot Water Heating Return	HWR

Air Conditioning Piping

Symbol Name	Symbol
Refrigerant Liquid	RL
Refrigerant Discharge	RD
Refrigerant Suction	RS
Condenser Water Supply	CWS
Condenser Water Return	CWR
Chilled Water Supply	CHWS
Chilled Water Return	CHWR
Makeup Water	MU
Humidification Line	H
Drain	D

INDUSTRIAL ELECTRICAL SYMBOLS...

Disconnect	Circuit Interrupter	Circuit Breaker with Thermal OL	Circuit Breaker with Magnetic OL	Circuit Breaker with Thermal and Magnetic Ol

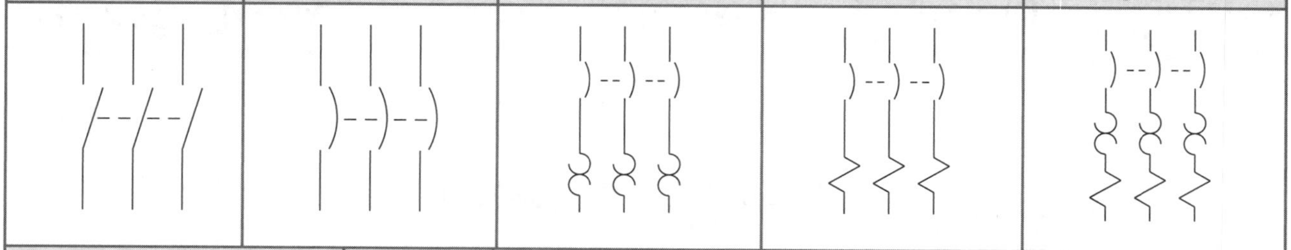

Limit Switches: Normally Open; Normally Closed; Held Closed; Held Open

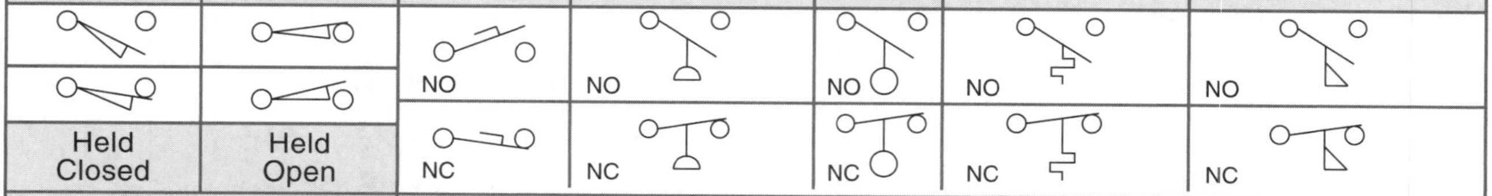

Foot Switches: NO; NC

Pressure and Vacuum Switches: NO; NC

Liquid Level Switch: NO; NC

Temperature-Actuated Switch: NO; NC

Flow Switch (Air, Water, Etc.): NO; NC

Speed (Plugging): F; F, R

Anti-plug: F, R

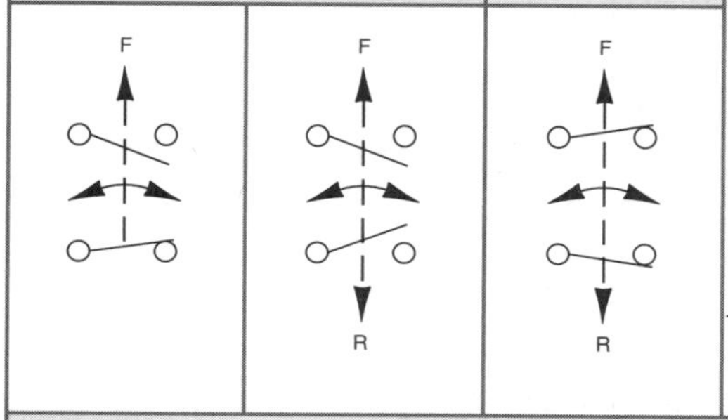

Symbols For Static Switching Control Devices

Static switching control is a method of switching electrical circuits without use of contacts, primarily by soild-state devices. Use symbols shown in table and enclose them in a diamond.

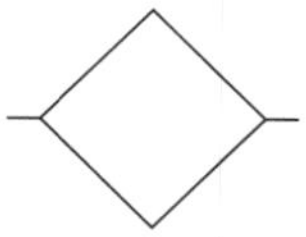

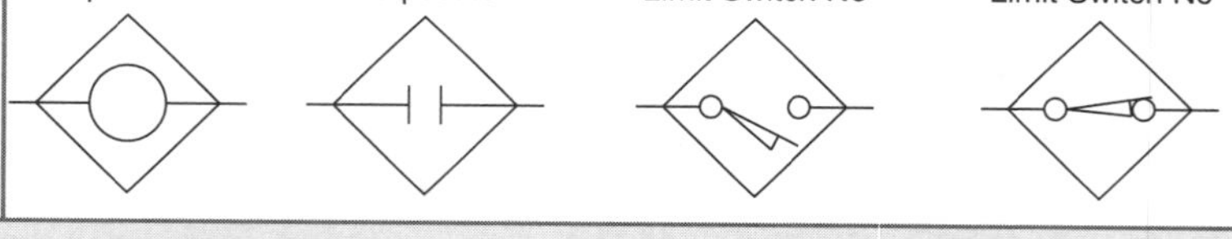

Input Coil — Output No — Limit Switch No — Limit Switch Nc

Selector

Two-position

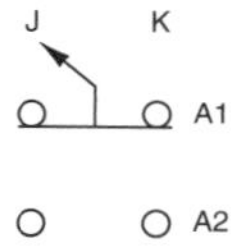

J K A1 A2

	J	K
A1	X	
A2		X

X-CONTACT CLOSED

Three-position

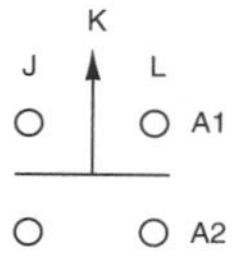

K J L A1 A2

	J	K	L
A1	X		
A2			X

X-CONTACT CLOSED

Two-position Selector Pushbutton

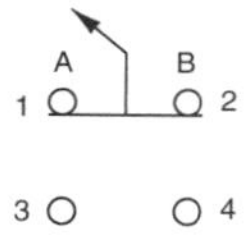

A B 1 2 3 4

CONTACTS	SELECTOR POSITION			
	A		B	
	BUTTON		BUTTON	
	FREE	DEPRESSED	FREE	DEPRESSED
1-2	X			
3-4		X	X	X

X - CONTACT CLOSED

Pushbuttons

Momentary Contact: Single Circuit (NO, NC); Double Circuit (NO AND NC); Mushroom Head; Wobble Stick

Maintained Contact: Two Single Circuit; One Double Circuit

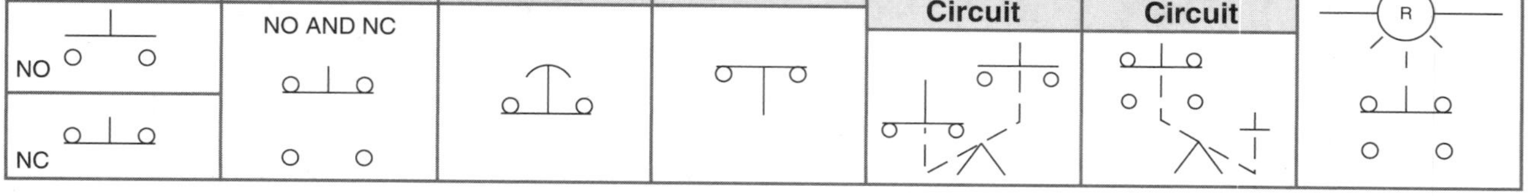

Illuminated: R

...INDUSTRIAL ELECTRICAL SYMBOLS...

Contacts								Overload Relays	
Instant Operating				Timed Contacts - Contact Action Retarded After Coil is:				Thermal	Magnetic
With Blowout		Without Blowout		Energized		De-energized			
NO	NC	NO	NC	NOTC	NCTO	NOTO	NCTC		

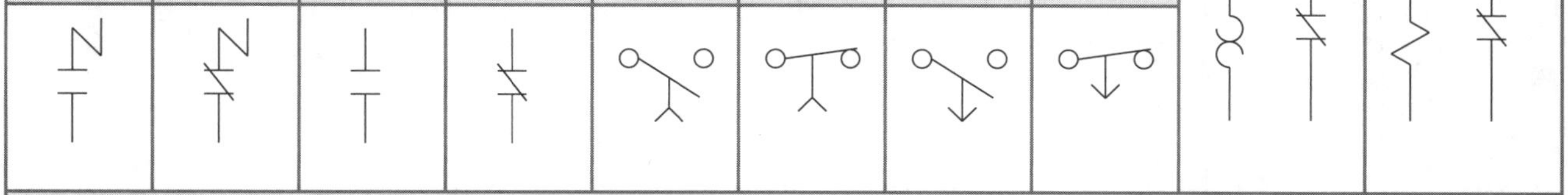

Supplementary Contact Symbols

Spst No		Spst Nc		Spdt		Terms
Single Break	Double Break	Single Break	Double Break	Single Break	Double Break	**SPST** Single-Pole, Single-Throw **SPDT** Single-Pole, Double-Throw
DPST, 2NO		**DPST, 2NC**		**DPDT**		**DPST** Double-Pole, Single-Throw
Single Break	Double Break	Single Break	Double Break	Single Break	Double Break	**DPDT** Double-Pole, Double-Throw **NO** Normally Open **NC** Normally Closed

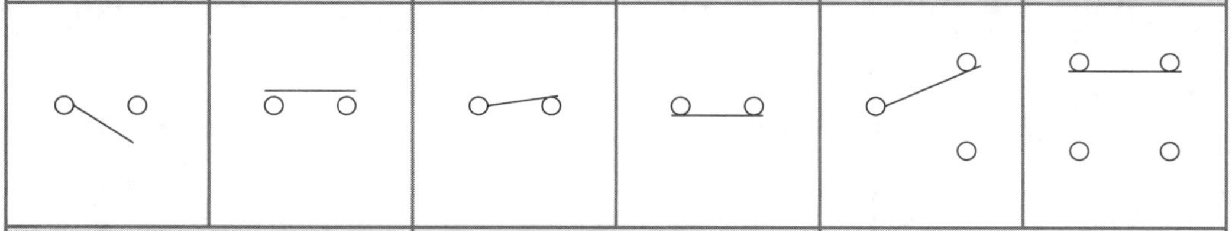

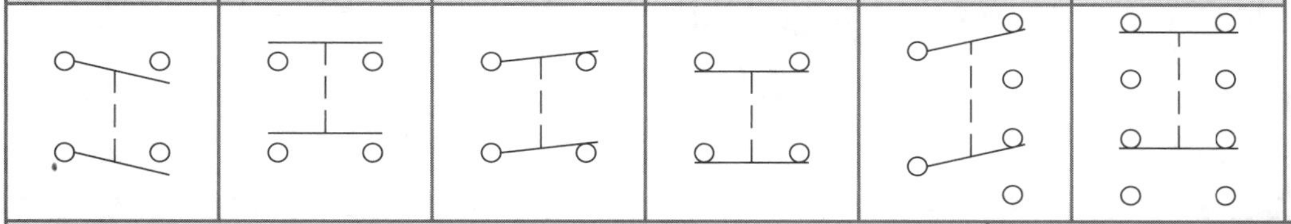

Meter (Instrument)

Indicate Type by Letter

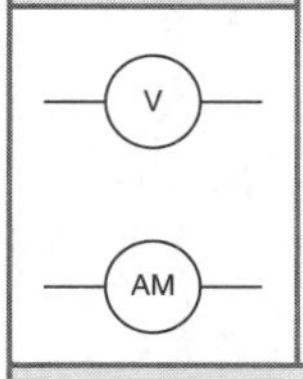

V

AM

To Indicate Function Of Meter Or Instrument, Place Specified Letter or Letters within Symbol

AM or A	Ammeter	VA	Voltmeter
AH	Ampere Hour	VAR	Varmeter
μA	Microammeter	VARH	Varhour Meter
mA	Millammeter	W	Wattmeter
PF	Power Factor	WH	Watthour Meter
V	Voltmeter		

Pilot Lights

Indicate Color By Letter

Non Push-to-test	Push-to-test
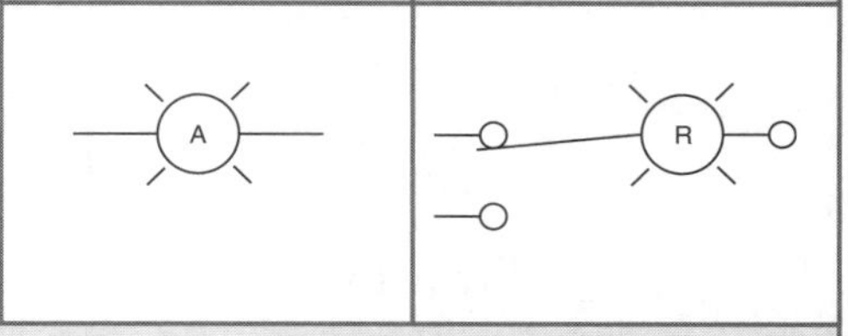 A	R

Inductors

Iron Core

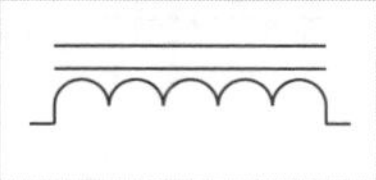

Air Core

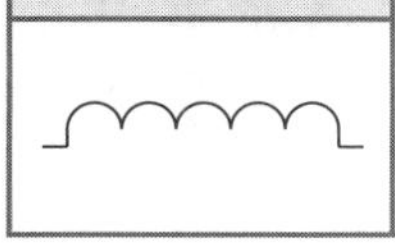

Coils

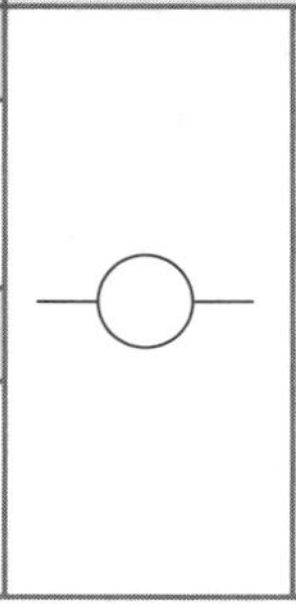

Dual-voltage Magnet Coils

High-voltage	Low-voltage	Blowout Coil
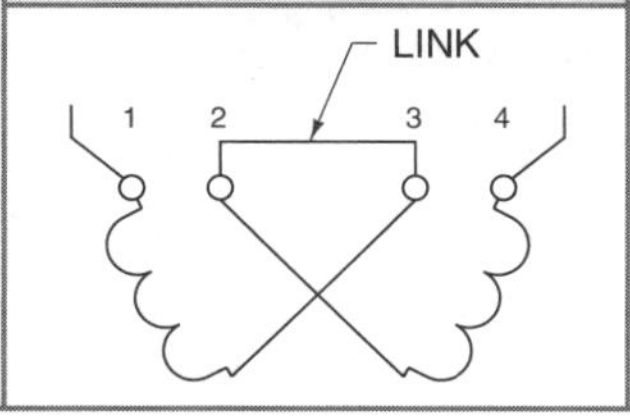 LINK 1 2 3 4	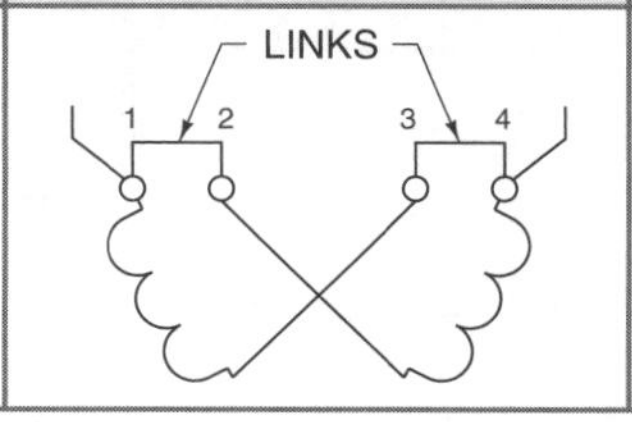 LINKS 1 2 3 4	

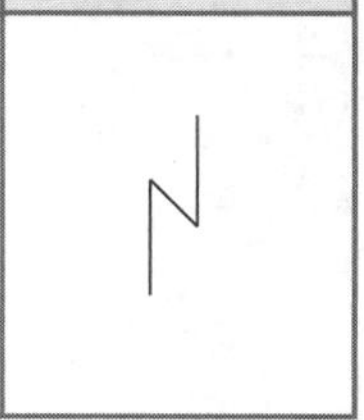

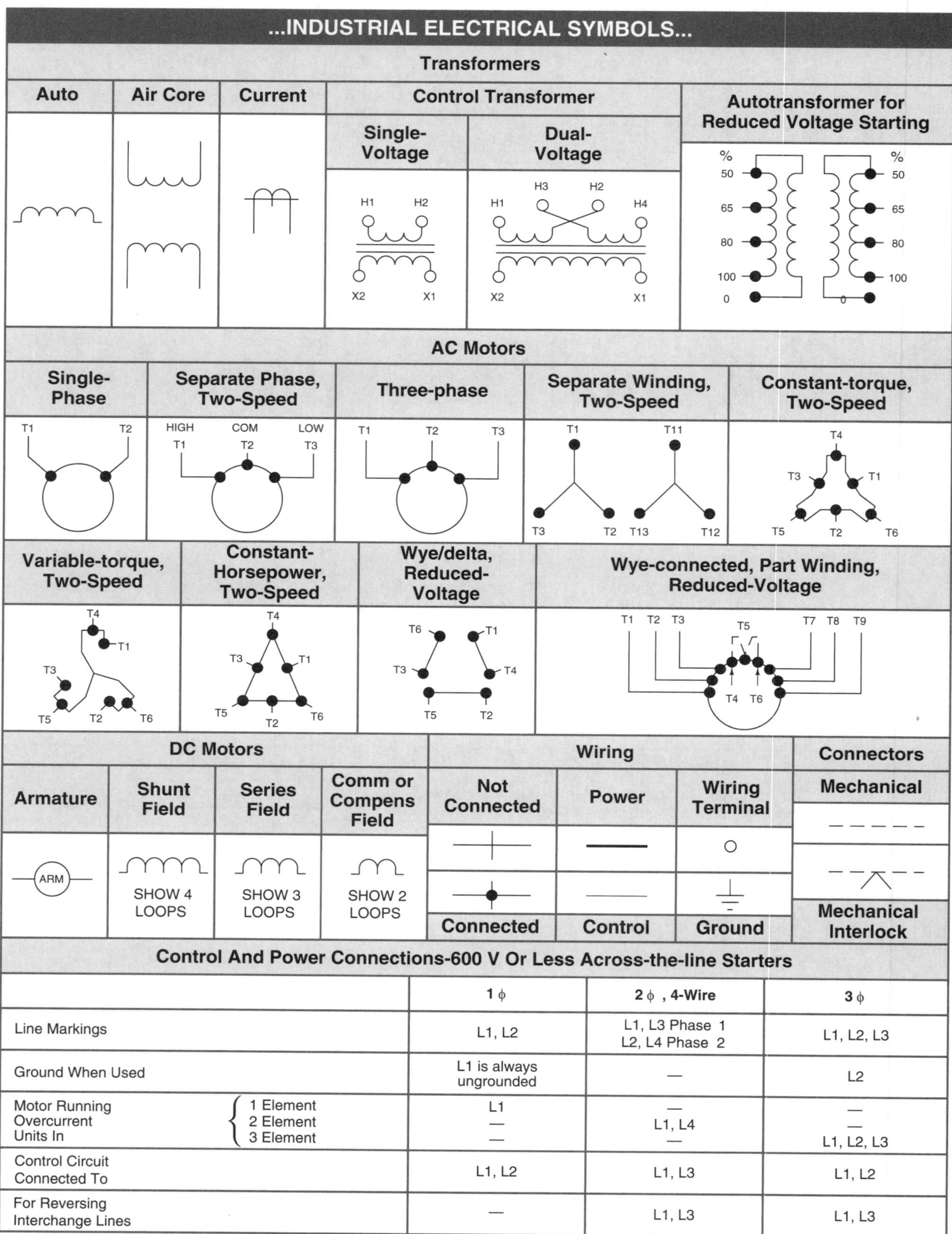

Control And Power Connections-600 V Or Less Across-the-line Starters

	1 ϕ	2 ϕ , 4-Wire	3 ϕ
Line Markings	L1, L2	L1, L3 Phase 1 L2, L4 Phase 2	L1, L2, L3
Ground When Used	L1 is always ungrounded	—	L2
Motor Running Overcurrent Units In { 1 Element 2 Element 3 Element	L1 — —	— L1, L4 —	— — L1, L2, L3
Control Circuit Connected To	L1, L2	L1, L3	L1, L2
For Reversing Interchange Lines	—	L1, L3	L1, L3

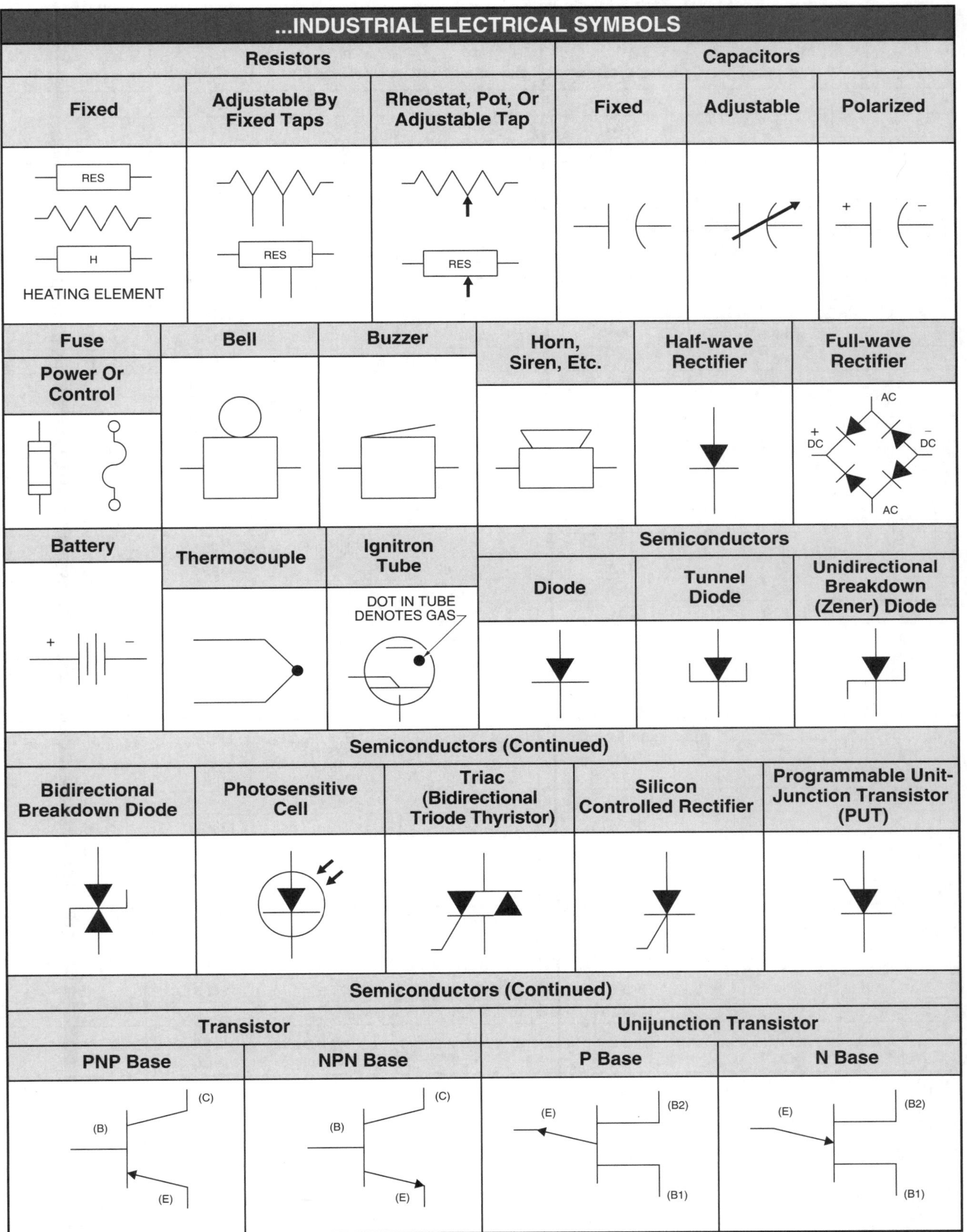
...INDUSTRIAL ELECTRICAL SYMBOLS
Resistors
Fixed
RES
H
HEATING ELEMENT
Adjustable By Fixed Taps
RES
Rheostat, Pot, Or Adjustable Tap
RES
Capacitors
Fixed
Adjustable
Polarized
+
−
Fuse
Power Or Control
Bell
Buzzer
Horn, Siren, Etc.
Half-wave Rectifier
Full-wave Rectifier
AC
+ DC
− DC
AC
Battery
+
−
Thermocouple
Ignitron Tube
DOT IN TUBE DENOTES GAS
Semiconductors
Diode
Tunnel Diode
Unidirectional Breakdown (Zener) Diode
Semiconductors (Continued)
Bidirectional Breakdown Diode
Photosensitive Cell
Triac (Bidirectional Triode Thyristor)
Silicon Controlled Rectifier
Programmable Unit-Junction Transistor (PUT)
Semiconductors (Continued)
Transistor
PNP Base
(B)
(C)
(E)
NPN Base
(B)
(C)
(E)
Unijunction Transistor
P Base
(E)
(B2)
(B1)
N Base
(E)
(B2)
(B1)

PROPERTIES OF SATURATED STEAM						
Gauge Pressure*	Absolute Pressure†	Temperature‡	Sensible Heat§	Latent Heat§	Total Heat§	Specific Volume Steam**
32	46.7	276.7	245.9	927.6	1173.5	9.08
34	48.7	279.4	248.5	925.8	1174.3	8.73
36	50.7	281.9	251.1	924.0	1175.1	8.40
38	52.7	284.4	253.7	922.1	1175.8	8.11
40	54.7	286.7	256.1	920.4	1176.5	7.83
42	56.7	289.0	258.5	918.6	1177.1	7.57
44	58.7	291.3	260.8	917.0	1177.8	7.33
46	60.7	293.5	263.0	915.4	1178.4	7.10
48	62.7	295.6	265.2	913.8	1179.0	6.89
50	64.7	297.7	267.4	912.2	1179.6	6.68
52	66.7	299.7	269.4	910.7	1180.1	6.50
54	68.7	301.7	271.5	909.2	1180.7	6.32
56	70.7	303.6	273.5	907.8	1181.3	6.16
58	72.7	305.5	275.3	906.5	1181.8	6.00
60	74.7	307.4	277.1	905.3	1182.4	5.84
62	76.7	309.2	279.0	904.0	1183.0	5.70
64	78.7	310.9	280.9	902.6	1183.5	5.56
66	80.7	312.7	282.8	901.2	1184.0	5.43
68	82.7	314.3	284.5	900.0	1184.5	5.31
70	84.7	316.0	286.2	898.8	1185.0	5.19
72	86.7	317.7	288.0	897.5	1185.5	5.08
74	88.7	319.3	289.4	896.5	1185.9	4.97
76	90.7	320.9	291.2	895.1	1186.3	4.87
78	92.7	322.4	292.9	893.9	1186.8	4.77
80	94.7	323.9	294.5	892.7	1187.2	4.67
82	96.7	325.5	296.1	891.5	1187.6	4.58
84	98.7	326.9	297.6	890.3	1187.9	4.49
86	100.7	328.4	299.1	889.2	1188.3	4.41
88	102.7	329.9	300.6	888.1	1188.7	4.33
90	104.7	331.2	302.1	887.0	1189.1	4.25
92	106.7	332.6	303.5	885.8	1189.3	4.17
94	108.7	333.9	304.9	884.8	1189.7	4.10
96	110.7	335.3	306.3	883.7	1190.0	4.03
98	112.7	336.6	307.7	882.6	1190.3	3.96
100	114.7	337.9	309.0	881.6	1190.6	3.90
102	116.7	339.2	310.3	880.6	1190.9	3.83
104	118.7	340.5	311.6	879.6	1191.2	3.77
106	120.7	341.7	313.0	878.5	1191.5	3.71
108	122.7	343.0	314.3	877.5	1191.8	3.65
110	124.7	344.2	315.5	876.5	1192.0	3.60
112	126.7	345.4	316.8	875.5	1192.3	3.54
114	128.7	346.5	318.0	874.5	1192.5	3.49

* in psig
† in psia
‡ in °F
§ in Btu/lb
** in cu ft/lb

continued

PROPERTIES OF SATURATED STEAM						
Gauge Pressure*	Absolute Pressure†	Temperature‡	Sensible Heat§	Latent Heat§	Total Heat§	Specific Volume Steam**
116	130.7	347.7	319.3	873.5	1192.8	3.44
118	132.7	348.9	320.5	872.5	1193.0	3.39
120	134.7	350.1	321.8	871.5	1193.3	3.34
125	139.7	352.8	324.7	869.3	1194.0	3.23
130	144.7	355.6	327.6	866.9	1194.5	3.12
135	149.7	358.3	330.6	864.5	1195.1	3.02
140	154.7	360.9	333.2	862.5	1195.7	2.93
145	159.7	363.5	335.9	860.3	1196.2	2.84
150	164.7	365.9	338.6	858.0	1196.6	2.76
155	169.7	368.3	341.1	856.0	1197.1	2.68
160	174.7	370.7	343.6	853.9	1197.5	2.61
165	179.7	372.9	346.1	851.8	1197.9	2.54
170	184.7	375.2	348.5	849.8	1198.3	2.48
175	189.7	377.5	350.9	847.9	1198.8	2.41
180	194.7	379.6	353.2	845.9	1199.1	2.35
185	199.7	381.6	355.4	844.1	1199.5	2.30
190	204.7	383.7	357.6	842.2	1199.8	2.24
195	209.7	385.7	359.9	840.2	1200.1	2.18
200	214.7	387.7	362.0	838.4	1200.4	2.14
210	224.7	391.7	366.2	834.8	1201.0	2.04
220	234.7	395.5	370.3	831.2	1201.5	1.96
230	244.7	399.1	374.2	827.8	1202.0	1.88
240	254.7	402.7	378.0	824.5	1202.5	1.81
250	264.7	406.1	381.7	821.2	1202.9	1.74
260	274.7	409.3	385.3	817.9	1203.2	1.68
270	284.7	412.5	388.8	814.8	1203.6	1.62
280	294.7	415.8	392.3	811.6	1203.9	1.57
290	304.7	418.8	395.7	808.5	1204.2	1.52
300	314.7	421.7	398.9	805.5	1204.4	1.47
310	324.7	424.7	402.1	802.6	1204.7	1.43
320	334.7	427.5	405.2	799.7	1204.9	1.39
330	344.7	430.3	408.3	796.7	1205.0	1.35
340	354.7	433.0	411.3	793.8	1205.1	1.31
350	364.7	435.7	414.3	791.0	1205.3	1.27
360	374.7	438.3	417.2	788.2	1205.4	1.24
370	384.7	440.8	420.0	785.4	1205.4	1.21
380	394.7	443.3	422.8	782.7	1205.5	1.18
390	404.7	445.7	425.6	779.9	1205.5	1.15
400	414.7	448.1	428.2	777.4	1205.6	1.12
420	434.7	452.8	433.4	772.2	1205.6	1.07
440	454.7	457.3	438.5	767.1	1205.6	1.02
460	474.7	461.7	443.4	762.1	1205.5	.98

* in psig
† in psia
‡ in °F
§ in Btu/lb
** in cu ft/lb

continued

PROPERTIES OF SATURATED STEAM

Gauge Pressure*	Absolute Pressure†	Temperature‡	Sensible Heat§	Latent Heat§	Total Heat§	Specific Volume Steam**
480	494.7	465.9	448.3	757.1	1205.4	.94
500	514.7	470.0	453.0	752.3	1205.3	.902
520	534.7	474.0	457.6	747.5	1205.1	.868
540	554.7	477.8	462.0	742.8	1204.8	.835
560	574.7	481.6	466.4	738.1	1204.5	.805
580	594.7	485.2	470.7	733.5	1204.2	.776
600	614.7	488.8	474.8	729.1	1203.9	.750
620	634.7	492.3	479.0	724.5	1203.5	.726
640	654.7	495.7	483.0	720.1	1203.1	.703
660	674.7	499.0	486.9	715.8	1202.7	.681
680	694.7	502.2	490.7	711.5	1202.2	.660
700	714.7	505.4	494.4	707.4	1201.8	.641
720	734.7	508.5	498.2	703.1	1201.3	.623
740	754.7	511.5	501.9	698.9	1200.8	.605
760	774.7	514.5	505.5	694.7	1200.2	.588
780	794.7	517.5	509.0	690.7	1199.7	.572
800	814.7	520.3	512.5	686.6	1199.1	.557

* in psig
† in psia
‡ in °F
§ in Btu/lb
** in cu ft/lb

Spirax Sarco, Inc.

REFRIGERATION SYMBOLS

Gauge

Sight Glass

High Side Float Valve

Low Side Float Valve

Immersion Cooling Unit

Cooling Tower

Natural Convection, Finned Type Evaporator

Forced Convection Evaporator

Pressure Switch

Hand Expansion Valve

Automatic Expansion Valve

Thermostatic Expansion Valve

Constant Pressure Valve, Suction

Thermal Bulb

Scale Trap

Self-contained Thermostat

Dryer

Filter And Strainer

Combination Strainer And Dryer

Evaporative Condensor

Heat Exchanger

Air-cooled Condensing Unit

Water-cooled Condensing Unit